THE CHEMICAL SOCIETY
MONOGRAPH FOR TEACHERS No. 8

Principles of Atomic Orbitals

(Revised SI edition)

N. N. GREENWOOD, DSc, ScD, FRIC

Professor of Inorganic and Structural Chemistry
University of Leeds

LONDON: THE CHEMICAL SOCIETY

Monographs for teachers

This is another publication in the series of Monographs for Teachers which was launched in 1959 by the Royal Institute of Chemistry. The initial aim of the series was to present concise and authoritative accounts of selected well-defined topics in chemistry for those who teach the subject at GCE Advanced level and above. This scope has now been widened to cover accounts of newer areas of chemistry or of interdisciplinary fields that make use of chemistry. Though intended primarily for teachers of chemistry, the monographs will doubtless be of value also to a wider readership, including students in further and higher education.

First published, 1964; Revised edition, 1968; Revised SI edition, August 1973. Reprinted, 1980

Published by The Chemical Society, Burlington House, London W1V 0BN, and distributed by The Chemical Society, Distribution Centre, Blackhorse Road, Letchworth, Herts. SG6 1HN

Printed by Adlard & Son Ltd., Bartholomew Press, Dorking

CONTENTS

1. Introduction

The correlation and interpretation of the experimental observations of chemistry depend ultimately on a knowledge of the atomic structure of the elements. In particular, it is important to know the spatial distribution of the outermost electrons and their relative energies of binding. With such information it is possible to understand the occurrence of covalent, ionic and metallic bonds; the relative stabilities of the various oxidation states of atoms; the stereochemistry of their compounds; magnetic properties; crystal structures; and the conditions under which compounds can deviate from ideal stoichiometry and exist over a range of composition. Quantum mechanics provides a framework for this detailed description of the electronic structure of atoms and leads to concepts such as atomic orbital, molecular orbital, hybridization, zero-point energy, spin–orbit coupling and ligand-field stabilization energy, which have become part of the language of modern chemistry.

A convenient starting point is the Schrödinger wave-equation, which takes into account the statistical nature of our knowledge of atoms and is an adequate description of electron behaviour as determined experimentally. Alternative formulations of the theory start from Heisenberg's uncertainty principle or adopt Dirac's symbolic method, but the mathematics and formalism in these presentations are less familiar to students of chemistry. Formal derivations of the results of wave-mechanics are available in several books[1–5] and will not be repeated here, but the results themselves, which are necessary for an understanding of chemistry, will be stated in order to illustrate the underlying concepts.

The Schrödinger equation involves both space and time coordinates, but for many purposes involving stationary states it is possible to separate out the time portion and concentrate on the time-independent equation. It is important to bear this initial simplification in mind for there are some problems (*eg* interaction with radiation, the phenomenon of magnetism, electron spin *etc*) which cannot be treated in this way. The time-independent Schrödinger equation for an electron in a one-electron atom is

$$\frac{\partial^2\psi}{\partial x^2}+\frac{\partial^2\psi}{\partial y^2}+\frac{\partial^2\psi}{\partial z^2}+\frac{8\pi^2 m_e}{h^2}(E-V)\psi = 0 \qquad 1$$

where x, y and z are the space coordinates, m_e is the mass of the electron, E its total energy, V its potential energy and h is Planck's constant. Equation (1) is a second-order partial differential equation; it is called an eigenvalue equation because the solutions, ψ, have significance only for certain values (eigenvalues) of the energy, E. For each value of E (*ie* for each atomic energy-level) there are one or more eigenfunctions, ψ, which are solutions to the equation. These functions must be finite, single-valued and continuous for physical significance and are then called wave-functions or orbitals. The incentive for solving the equation is that ψ serves as a description of the electron and gives information *inter alia* about the spatial distribution, angular momentum and energy of an electron in that orbital. The functions, ψ, themselves have no physical reality; it is $\psi^2 \mathrm{d}v$ which can be correlated with reality by giving the probability of finding an electron of energy, E, in the element of volume, $\mathrm{d}v$. The wave-equation and its solutions, ψ, are best thought of as mathematical devices for calculating this probability.

2. The Hydrogen Atom

Solution of the wave-equation

Precise solutions of the time-independent Schrödinger equation have been obtained only for those systems in which one electron is bound to a nucleus of charge Ze (where $Z = 1$ for H, 2 for He^+, 3 for Li^{2+}, *etc.*). When more than one electron is present, the interelectron repulsion terms in the potential-energy function, V, necessitate approximate methods of solution. Although chemists are usually interested in more complex systems than the hydrogen atom, it is important to study this example in some detail because many of the results can be taken over directly for the more complicated systems and many of the fundamental concepts emerge at this stage.

The potential energy, V, of an electron, charge $-e$ at a distance r from a nucleus of charge $+Ze$ is $-Ze^2/(4\pi\epsilon_0 r)$, so that the time-independent Schrödinger equation for a hydrogen-like atom becomes

$$\nabla^2\psi + \frac{8\pi^2 m_e}{h^2}\left(E + \frac{Ze^2}{(4\pi\epsilon_0 r)}\right)\psi = 0 \qquad 2$$

where the operator ∇^2 has been used to denote $(\partial^2/\partial x^2)+(\partial^2/\partial y^2)+(\partial^2/\partial z^2)$. This equation is best solved by transforming from Cartesian coordinates, x, y, z, to spherical polar coordinates, r, θ, ϕ. The relation between these coordinates is shown in *Fig. 1*, from which it is clear that

$$z = r\cos\theta$$

$$x = r\sin\theta\cos\phi$$

$$y = r\sin\theta\sin\phi$$

$$r^2 = x^2+y^2+z^2$$

$$\cos\theta = z/r$$

$$\tan\phi = y/x$$

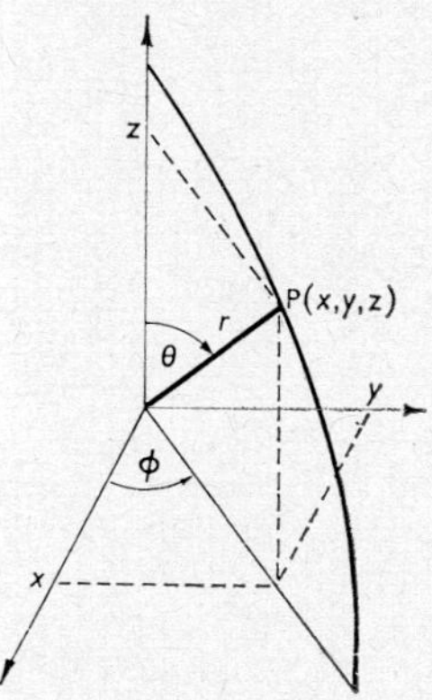

FIG. 1. Spherical polar coordinates

The radial distance coordinate, r, has values from 0 to ∞, the angle θ has values from 0 to π and the angle ϕ from 0 to 2π.

Equation (2) can now be solved by separating the variables r, θ

and ϕ, that is by looking for solutions, $\psi(r,\theta,\phi)$, which are the product of three functions, one of which, $R(r)$, depends on r alone; one, $\Theta(\theta)$, which depends on θ alone; and one, $\Phi(\phi)$, which depends on ϕ alone:

$$\psi(r,\theta,\phi) = R(r)\Theta(\theta)\Phi(\phi) \qquad 3$$

This leads to three simpler equations each in one variable only, which can then be solved and the solutions multiplied together to give the wave-function, ψ. The first of these simpler equations has the form

$$\frac{\partial^2\Phi(\phi)}{\partial\phi^2}+m^2\Phi(\phi) = 0 \qquad 4$$

One solution of this equation is $\Phi(\phi) = \sin m\phi$. This holds for any numerical value of the constant, m, but the solution will only be acceptable for our purpose if it is finite, single-valued and continuous. A sine function is finite and continuous but it is single-valued only if m is zero or a whole number, because ϕ runs from 0 to 2π and sin $(m\times 0)$ must equal sin $(m\times 2\pi)$. We see, therefore, that the first quantum number, m, arises from the condition that the wave-function must be single-valued if it is to have physical significance. It is one of the most satisfying features of wave-mechanics that the quantum numbers arise quite naturally from the mathematics and are not imposed arbitrarily on the system as in the older quantum theories of Bohr and Sommerfeld.

A simple example of the quantum condition occurs in the vibration of an elastic string which is fixed at both ends (boundary conditions). Physically, as the string has a finite length and the number of nodes must be integral, only certain solutions are acceptable, as illustrated in *Fig. 2*. In this example, the continuity requirement at 0 and x is not relevant, as the two ends of the string do not have to join up.

The general solution of equation (4) is

$$\Phi_m(\phi) = ce^{\pm im\phi} \qquad 5$$

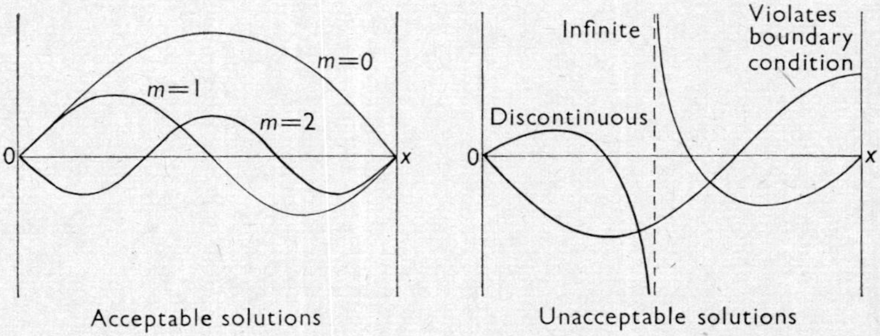

FIG. 2. Vibrations of an elastic string

where c is a constant to be determined by normalization (*see* Appendix 1, p. 42) and $i = \sqrt{(-1)}$. The solutions thus contain an imaginary part, and it is sometimes convenient to combine the solutions in pairs in order to obtain real forms of the solutions. To do this we use the relations

$$\cos|m|\phi = (e^{im\phi}+e^{-im\phi})/2; \quad \sin|m|\phi = -(e^{im\phi}-e^{-im\phi})/2i \qquad 6$$

where $|m|$ is the numerical value of m independent of sign.

The solving of the other two separated equations to find the functions $\Theta(\theta)$ and $R(r)$ is much more complicated and for our purpose it is only necessary to appreciate that each equation introduces a further quantum number, l or n, which must be integral for physical significance. The wave-function or orbital can then be written

$$\psi(r,\theta,\phi) = R_{n,l}(r)\Theta_{l,m}(\theta)\Phi_m(\phi) = R_{n,l}(r)A_{l,m}(\theta,\phi) \qquad 7$$

Equation (7) states that an orbital can be represented as the product of three functions, one of which depends only on the distance, r, from the nucleus and is determined by the quantum numbers n and l, and the other two of which depend only on the direction, θ,ϕ, and are determined by the quantum numbers l and m. This separation of the orbital into a radial function $R(r)$ and an angular function $A(\theta,\phi)$ is of fundamental importance.

It is also evident from the mathematics that $|m|$ can never be greater than l and that l is always less than n (*see* Appendix 1, p. 42).

The quantum number, n, is called the principal quantum number and defines the sequence of energy levels or shells within the atom. For a hydrogen-like atom, the energy of an electron in an orbital of principal quantum number n is $E = C/n^2$ where $C = -2\pi^2 m_e Z^2 e^4/\{(4\pi\epsilon_0)^2 h^2\}$ and m_e is the mass of the electron. The azimuthal or orbital quantum number, l, defines the shape of the orbital, and orbitals having values of l equal to 0, 1, 2 and 3 are designated s, p, d and f orbitals. The orbital quantum number also determines the orbital angular momentum:

$$M = \sqrt{l(l+1)}(h/2\pi)$$

The magnetic quantum number, m, gives the component of the orbital angular momentum in the direction of the z-axis: $M_z = m(h/2\pi)$. It can be regarded as specifying the orientation of the orbital with respect to this axis, which is defined by an applied magnetic field. As m can range from $+l$ through zero to $-l$, it can take on $(2l+1)$ values, as shown in *Fig. 3* for the case of the five d orbitals ($l = 2$). In an isolated atom, orbitals which have the same value for n and l normally have the same energy but this degeneracy is removed in the presence of a magnetic field.

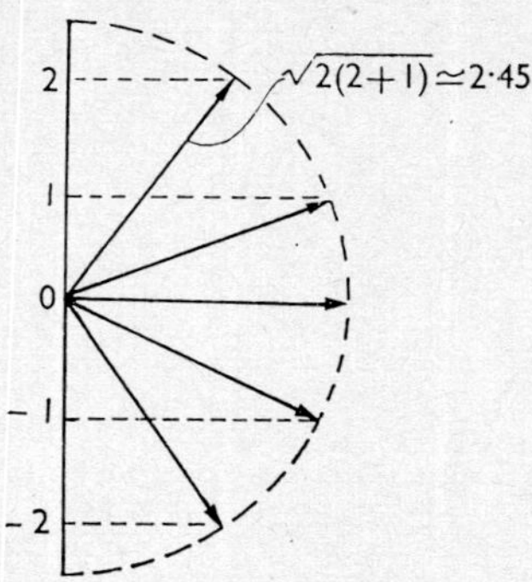

FIG. 3. Magnetic quantum numbers (m) for d orbitals ($l = 2$)

It is interesting to note that the maximum value of M_z is $l(h/2\pi)$ which is less than the total orbital angular momentum $\sqrt{l(l+1)}(h/2\pi)$. This is a consequence of the uncertainty principle because, if the z-component equalled the total value, then the x- and y-components would be known to be zero and one would thus have simultaneously precise knowledge both of the magnitude of the angular momentum and of the direction of the axis about which the motion was taking place.

Radial functions $R_{n,l}(r)$

The radial functions for a hydrogen-like atom consist of a constant, multiplied by a simple polynomial in r and by a function which decreases exponentially with increase in r. This exponential decay is more gradual for larger n and more rapid for larger Z (*see* Appendix 1, eqn 5, p. 43). The functions are given in Table 1 for the first ten combinations of n and l. The graphs for some of these functions are plotted in *Fig. 4* which shows that the radial functions for 1s, 2p, 3d and 4f orbitals have no nodes and are everywhere of the same sign, *eg* positive. In general, $R_{n,l}(r)$ becomes zero $(n-l-1)$ times between 0 and ∞ as the degree of the associated Laguerre polynomial is $(n+l)-(2l+1) = (n-l-1)$ (*see* Appendix 1, p. 43). It is also clear from the graphs that all the functions approach zero exponentially for large values of r and that only s orbitals ($l = 0$) have a finite value of $R_{n,l}(r)$ at the nucleus. The graphs have all been plotted with the same horizontal scale (r in units of $a_0 = 52.9$ pm) but the scale of the vertical axis has been varied to give plots of convenient size. For example, the vertical scale for the 1s radial function is only one hundredth of that used for the 4f function.

The importance of the radial functions, which themselves have no physical reality, is that they can be used to calculate (among other

Table 1. Normalized radial functions $R_{n,l}(r)$ for hydrogen-like atoms*

$$R_{n,l}(r) = -\sqrt{\frac{4(n-l-1)!Z^3}{n^4[(n+l)!]^3}} \times \left(\frac{2Zr}{n}\right)^l L_{n+1}^{2l+1}\left(\frac{2Zr}{n}\right) \times e^{-Zr/n}$$

Orbital	*n*	*l*	$R_{n,l}$ =	*Constant* ×	*Polynomial* ×	*Expon.*
1s	1	0	$R_{1,0}$	$2Z^{3/2}$	1	e^{-Zr}
2s	2	0	$R_{2,0}$	$\frac{Z^{3/2}}{2\sqrt{2}}$	$(2-Zr)$	$e^{-Zr/2}$
2p	2	1	$R_{2,1}$	$\frac{Z^{3/2}}{2\sqrt{6}}$	Zr	$e^{-Zr/2}$
3s	3	0	$R_{3,0}$	$\frac{2Z^{3/2}}{81\sqrt{3}}$	$(27-18Zr+2Z^2r^2)$	$e^{-Zr/3}$
3p	3	1	$R_{3,1}$	$\frac{4Z^{3/2}}{81\sqrt{6}}$	$(6Zr-Z^2r^2)$	$e^{-Zr/3}$
3d	3	2	$R_{3,2}$	$\frac{4Z^{3/2}}{81\sqrt{30}}$	Z^2r^2	$e^{-Zr/3}$
4s	4	0	$R_{4,0}$	$\frac{Z^{3/2}}{768}$	$(192-144Zr+ 24Z^2r^2-Z^3r^3)$	$e^{-Zr/4}$
4p	4	1	$R_{4,1}$	$\frac{Z^{3/2}}{256\sqrt{15}}$	$(80Zr-20Z^2r^2+ Z^3r^3)$	$e^{-Zr/4}$
4d	4	2	$R_{4,2}$	$\frac{Z^{3/2}}{768\sqrt{5}}$	$(12Z^2r^2-Z^3r^3)$	$e^{-Zr/4}$
4f	4	3	$R_{4,3}$	$\frac{Z^{3/2}}{768\sqrt{35}}$	Z^3r^3	$e^{-Zr/4}$

* These functions are discussed more fully in Appendix 1 (p. 42) which also defines the associated Laguerre polynomial, L. In this table the 'constant' incorporates the constant term and sign of the worked polynomial.

things) the probability of an electron being at a certain distance, r, from the nucleus. This probability is $R^2_{n,l}(r)dv$, where dv, the volume of a spherical shell distance r from the nucleus, is $4\pi r^2 dr$. The physically significant quantity is therefore $4\pi R^2_{n,l}(r)\, r^2 dr$, and this is called the radial distribution function.

Plots of $4\pi R^2_{n,l}(r)\, r^2 dr$ are shown in *Fig. 4*, from which it is evident that the probability of finding an electron is always positive and that it drops to zero at the nucleus even for electrons in s orbitals. This is because r^2 is zero at the nucleus even though $R^2_{n,l}(r)$ is finite for these orbitals. The graphs also show that the radius at which there is maximum probability of finding the electron increases very rapidly with increase in the principal quantum number, n, the maxima occurring at r = 1, 4, 9 and 16 (*ie* n^2) for 1s, 2p, 3d and 4f orbitals. In making these comparisons we must remember that the actual

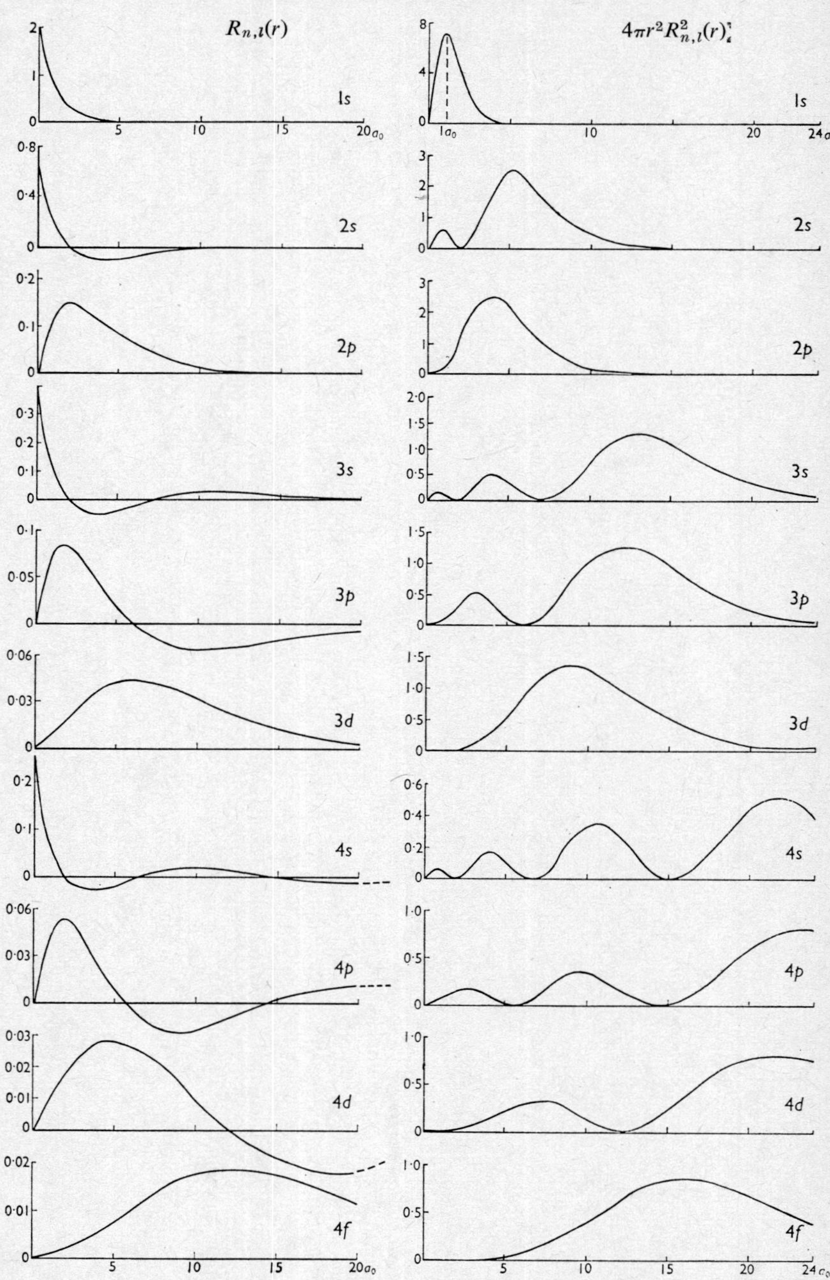

FIG. 4. Radial functions for a hydrogen atom

volume, dv, of the spherical shell considered is not constant but increases with r.

Another point that emerges from *Fig. 4* is that the radius of maximum probability decreases slightly with increase in the azimuthal quantum number, l. Of more importance chemically, however, is the fact that, at points very close to the nucleus, the values of $R_{n,l}(r)$ (and hence of $R^2_{n,l}(r)\, r^2 \mathrm{d}r$) are greater for s orbitals than for p orbitals and these in turn are larger than for d orbitals. This follows at once from the equations in Table 1, in which the main difference between $R_{n,l}(r)$ for a 3s, 3p and 3d orbital, for example, is the polynomial in r. For very small values of r the polynomial is approximately a constant ($= 27$) for the 3s orbital, but it depends on r for the 3p orbital and on r^2 for the 3d orbital. Hence the probability of finding an electron of principal quantum number 3 in a spherical shell close to the nucleus is greatest for a 3s orbital and least for a 3d orbital. The significance of this in many-electron atoms is discussed more fully on p. 22.

In the preceding paragraphs we have considered the probability of finding an electron in a spherical shell at a distance r from the nucleus, the probability being summed over all possible angles. However, the probability of finding an electron frequently depends on the particular direction chosen, *ie* the probability may not be evenly distributed over the spherical shell, but may be greater in some directions than in others with respect to some externally fixed direction imposed on the atom, for example by means of a magnetic field. Information about this variation can be obtained from the angular function, which comprises the second part of the orbital (eqn 7, p. 5). The possibility of non-spherically symmetrical electron distributions is of prime importance in chemistry because, if the probability of finding an electron in one particular direction is greater than in another, then presumably the atom will form a bond preferentially in that direction and we have a basis for interpreting the stereochemistry of covalent compounds. Once again we shall use the hydrogen atom as the starting point but it will be shown on p. 18 that the results also hold without modification for all other atoms.

Angular functions $A_{l,m}(\theta,\phi) = \Theta_{l,m}(\theta)\, \Phi_m(\phi)$

The angular functions, as their name implies, depend only on direction and are independent of the distance, r, from the nucleus. The functions $\Theta_{l,m}(\theta)$ consist of a constant and a simple polynomial in $\cos\theta$ and $\sin\theta$ (*see* Appendix 1, p. 43). Table 2 summarizes these functions for the first ten combinations of l and m. The functions $\Phi_m(\phi)$ are simply $\mathrm{e}^{\mathrm{i}m\phi}/\sqrt{(2\pi)}$, which gives $1/\sqrt{(2\pi)}$ for $m = 0$,

Table 2. Normalized functions, $\Theta_{l,m}(\theta)$

$$\Theta_{l,m}(\theta) = \sqrt{\frac{(2l+1)(l-|m|)!}{2(l+|m|)!}} \times P_l^{|m|}(\cos\theta)$$

Orbital	*l*	*m*	$\Theta_{l,m}$	= *Constant*	× *Polynomial*
s	0	0	$\Theta_{0,0}$	$\frac{1}{2}\sqrt{2}$	$\times 1$
p_z	1	0	$\Theta_{1,0}$	$\frac{1}{2}\sqrt{6}$	$\times \cos\theta$
p_x, p_y	1	± 1	$\Theta_{1,\pm 1}$	$\frac{1}{2}\sqrt{3}$	$\times \sin\theta$
d_{z^2}	2	0	$\Theta_{2,0}$	$\frac{1}{4}\sqrt{10}$	$\times (3\cos^2\theta - 1)$
d_{zx}, d_{zy}	2	± 1	$\Theta_{2,\pm 1}$	$\frac{1}{2}\sqrt{15}$	$\times \sin\theta\cos\theta$
$d_{x^2-y^2}$, d_{xy}	2	± 2	$\Theta_{2,\pm 2}$	$\frac{1}{4}\sqrt{15}$	$\times \sin^2\theta$
f_{z^3}	3	0	$\Theta_{3,0}$	$\frac{1}{4}\sqrt{14}$	$\times (5\cos^3\theta - 3\cos\theta)$
f_{z^2x}, f_{z^2y}	3	± 1	$\Theta_{3,\pm 1}$	$\frac{1}{8}\sqrt{42}$	$\times \sin\theta\,(5\cos^2\theta - 1)$
$f_{z(x^2-y^2)}$, f_{zxy}	3	± 2	$\Theta_{3,\pm 2}$	$\frac{1}{4}\sqrt{105}$	$\times \sin^2\theta\cos\theta$
f_{x^3}, f_{y^3}	3	± 3	$\Theta_{3,\pm 3}$	$\frac{1}{8}\sqrt{70}$	$\times \sin^3\theta$

$e^{\pm i\phi}/\sqrt{(2\pi)}$ for $m = \pm 1$, $e^{\pm 2i\phi}/\sqrt{(2\pi)}$ for $m = \pm 2$, *etc.* By means of the relations (6) on p. 5 these imaginary functions can be combined in pairs to give the corresponding real functions listed in Table 3.

The total angular dependence functions, $A_{l,m}(\theta,\phi)$, are the products of the separate Θ and Φ functions and are given in Table 4. It is essential to have a clear mental picture of the shapes and symmetry properties of these composite functions. They can be plotted as polar diagrams and the figures so obtained are illustrated in *Fig. 5*. The angular part of the total wave-function, ψ, and its graphical representation on a polar diagram have been the subject of more misrepresentation than any other concept in wave-mechanics (except perhaps the concept of resonance), and it is worth while to consider in some detail the precise interpretation of these functions, and how far they can be used to give a pictorial representation of the electron distribution within an atom. It should be stressed that the diagrams do not represent the orbitals, ψ, themselves but only the angular dependence of ψ. To obtain the complete orbital the angular function, $A_{l,m}(\theta,\phi)$, must be multiplied by the radial function, $R_{l,n}(r)$.

Table 3. Real forms of the normalized functions, $\Phi_m(\phi)$

$$\Phi_0(\phi) = 1/\sqrt{(2\pi)}$$

$$\Phi_{\pm 1}(\phi) = \begin{cases} (1/\sqrt{\pi})\cos\phi \\ (1/\sqrt{\pi})\sin\phi \end{cases}$$

$$\Phi_{\pm 2}(\phi) = \begin{cases} (1/\sqrt{\pi})\cos 2\phi = (1/\sqrt{\pi})(2\cos^2\phi - 1) \\ (1/\sqrt{\pi})\sin 2\phi = (1/\sqrt{\pi})(2\sin\phi\cos\phi) \end{cases}$$

$$\Phi_{\pm 3}(\phi) = \begin{cases} (1/\sqrt{\pi})\cos 3\phi = (1/\sqrt{\pi})(4\cos^3\phi - 3\cos\phi) \\ (1/\sqrt{\pi})\sin 3\phi = (1/\sqrt{\pi})(3\sin\phi - 4\sin^3\phi) \end{cases}$$

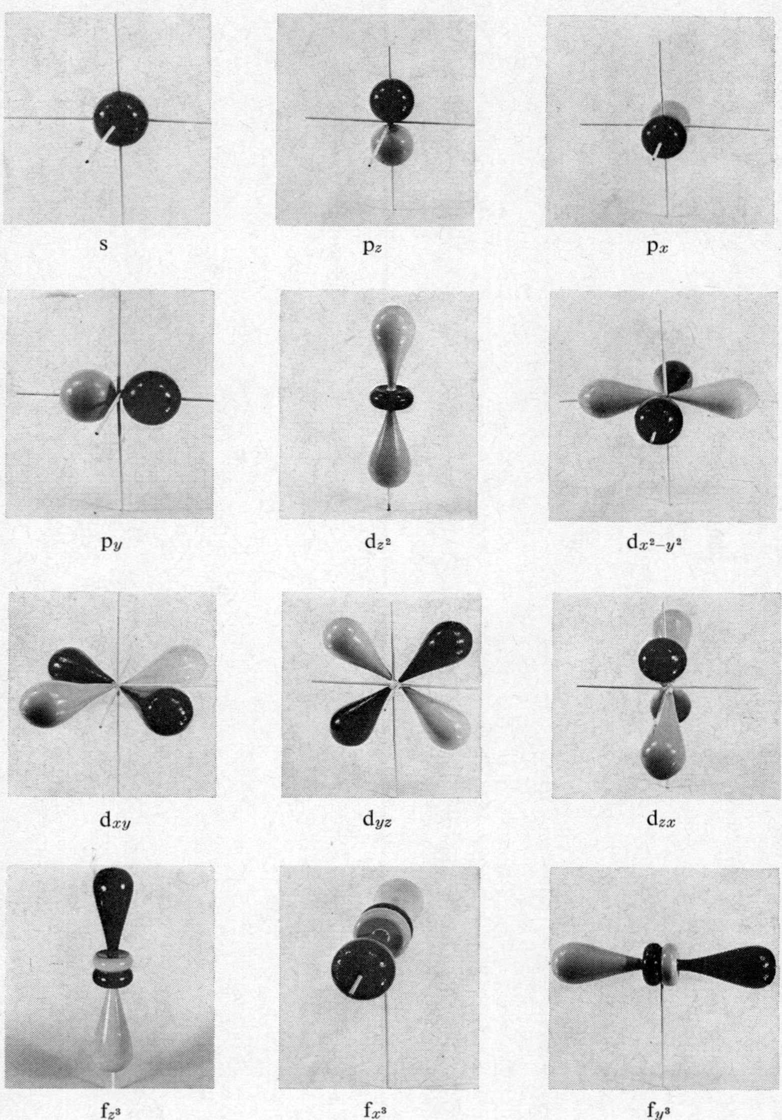

FIG. 5. Angular dependence functions (*continued overleaf*)

FIG. 5 (*cont.*)

For s orbitals ($l = 0$) the angular dependence function, A, is constant; *ie* the value of the function, $1/(2\sqrt{\pi})$, is independent of both θ and ϕ. This is represented on the polar diagram as a sphere of radius $1/(2\sqrt{\pi})$ since all points on the surface of a sphere are equidistant from its centre (the nucleus). It follows that the variation of ψ depends only on the radial function, R, so that the orbital is spherically symmetrical. In other words, the hydrogen atom is spherically symmetrical in its ground state, a property not possessed by the Bohr model, which was planar.

For p orbitals ($l = 1$) the polar diagram of the angular dependence function comprises two spheres in contact, each of diameter $\sqrt{3}/(2\sqrt{\pi})$, one being positive and the other negative (*see Fig. 5*). Again, the interpretation of the diagram is that the magnitude of the function $A_{l,m}(\theta,\phi)$ in a given direction, θ,ϕ, is equal to the length of the line drawn from the origin to the surface in that direction. The designation of the three p orbitals as p_x, p_y and p_z follows from the expressions for the angular functions in Table 4 which are identical with those used on p. 3 to define the polar coordinates r, θ, ϕ in terms of the Cartesian coordinates x, y, z. Table 4 also shows that $A(p_z)$, *ie* $A_{1,0}$, is independent of ϕ and hence cylindrically symmetrical about the z-axis. A section through the polar diagram in the xz plane ($\phi = 0$) consists of two contiguous circles (*Fig. 6a*). This is

Table 4. Normalized angular dependence functions, $A_{l,m}(\theta,\phi) = \Theta_{l,m}(\theta)\Phi_m(\phi)$

Orbital	*Angular dependence function*
s	$\frac{1}{2\sqrt{\pi}}$
p_z	$\frac{\sqrt{3}}{2\sqrt{\pi}} \cos\theta$
p_x	$\frac{\sqrt{3}}{2\sqrt{\pi}} \sin\theta \cos\phi$
p_y	$\frac{\sqrt{3}}{2\sqrt{\pi}} \sin\theta \sin\phi$
d_{z^2}	$\frac{\sqrt{5}}{4\sqrt{\pi}} (3\cos^2\theta - 1)$
d_{zx}	$\frac{\sqrt{15}}{2\sqrt{\pi}} \cos\theta \sin\theta \cos\phi$
d_{zy}	$\frac{\sqrt{15}}{2\sqrt{\pi}} \cos\theta \sin\theta \sin\phi$
$d_{x^2-y^2}$	$\frac{\sqrt{15}}{4\sqrt{\pi}} \sin^2\theta\,(2\cos^2\phi - 1)$
d_{xy}	$\frac{\sqrt{15}}{2\sqrt{\pi}} \sin^2\theta \sin\phi \cos\phi$
f_{z^3}	$\frac{\sqrt{7}}{4\sqrt{\pi}} (5\cos^3\theta - 3\cos\theta)$
f_{z^2x}	$\frac{\sqrt{42}}{8\sqrt{\pi}} (5\cos^2\theta - 1) \sin\theta \cos\phi$
f_{z^2y}	$\frac{\sqrt{42}}{8\sqrt{\pi}} (5\cos^2\theta - 1) \sin\theta \sin\phi$
$f_{z(x^2-y^2)}$	$\frac{\sqrt{105}}{4\sqrt{\pi}} \cos\theta \sin^2\theta\,(2\cos^2\phi - 1)$
f_{zxy}	$\frac{\sqrt{105}}{2\sqrt{\pi}} \cos\theta \sin^2\theta \cos\phi \sin\phi$
f_{x^3}	$\frac{\sqrt{70}}{8\sqrt{\pi}} \sin^3\theta\,(4\cos^3\phi - 3\cos\phi)$
f_{y^3}	$\frac{\sqrt{70}}{8\sqrt{\pi}} \sin^3\theta\,(3\sin\phi - 4\sin^3\phi)$

less familiar as a plot of $\cos\theta$ than the Cartesian plot shown in *Fig. 6b*, but it can readily be seen to be the shape generated by a line of length $\frac{1}{2}\sqrt{(3/\pi)}\cos\theta$ since the diameter of a circle subtends an angle of 90° at its circumference so that the length of the line A is $\frac{1}{2}\sqrt{(3/\pi)}\cos\theta$. Along the z-axis θ is zero, $\cos\theta$ is 1 and the function has its maximum value. As θ increases $\cos\theta$ diminishes, and the length of the line from the origin to the circumference also diminishes until, when θ is 90°, *ie* along the x-axis, the function is zero. This means that there is zero probability of finding a p_z electron in this direction. As θ increases still further, $\cos\theta$ becomes negative and

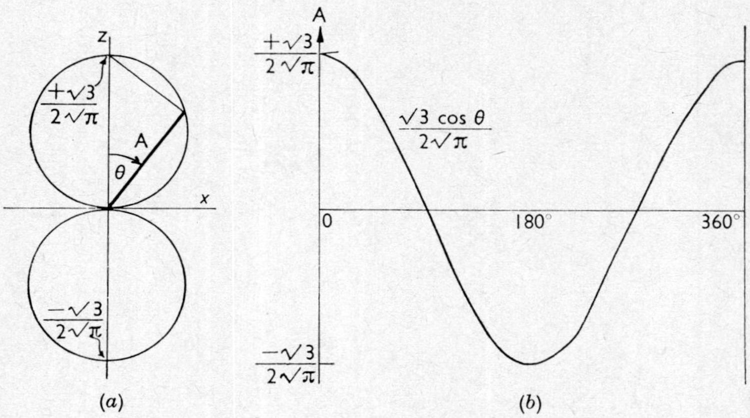

FIG. 6. Polar and Cartesian plots of $\sqrt{3}\cos\theta/(2\sqrt{\pi})$

the lower circle carries a negative sign. At 180°, *ie* in the $-z$ direction, $\cos\theta$ is -1 and the function has its maximum negative value. Further increase in θ diminishes the value of the function until at 270°, $(-x)$-axis, it is again zero. In the final quadrant, the length of the line from the origin to the curve increases until at $\theta = 360°$ it has returned to the original value at $\theta = 0°$. The polar diagrams for $A(p_x)$ and $A(p_y)$ have identical shapes to that of $A(p_z)$ but are oriented with their maximum extensions along the x- and y-axes, respectively.

The angular dependence functions themselves have no physical reality and cannot be observed directly by experiment. The physically significant quantity is $A^2_{l,m}(\theta,\phi)$. For each particular orbital, A^2 gives the probability of finding an electron in the direction θ,ϕ at any distance from the nucleus to infinity. We saw on p. 7 that $R^2_{n,l}(r)$ was the probability of finding an electron at a distance r from the nucleus in any direction. Hence, the probability of finding an electron simultaneously at a distance r and in the direction θ,ϕ is $R^2_{n,l}(r)\,A^2_{l,m}(\theta,\phi)$ which equals $\psi^2_{n,l,m}(r,\theta,\phi)$.

The directional properties of the various d orbitals are more complicated but can be considered in the same way. The nomenclature d_{z^2}, d_{xz}, *etc* is derived from the trigonometric functions defining A in Table 4 and the equations for transforming Cartesian to polar coordinates on p. 3. The polar diagram of d_{z^2} (*see Fig. 5*) consists of two positive lobes directed along the z- and $(-z)$-axes and separated from a negative region by two conical nodes at $\theta = 54.73°$ and 125.27°. This follows from Table 4 which gives $A(d_{z^2}) = \frac{1}{4}\sqrt{(5/\pi)}$ $(3\cos^2\theta-1)$. The function is independent of ϕ and is therefore

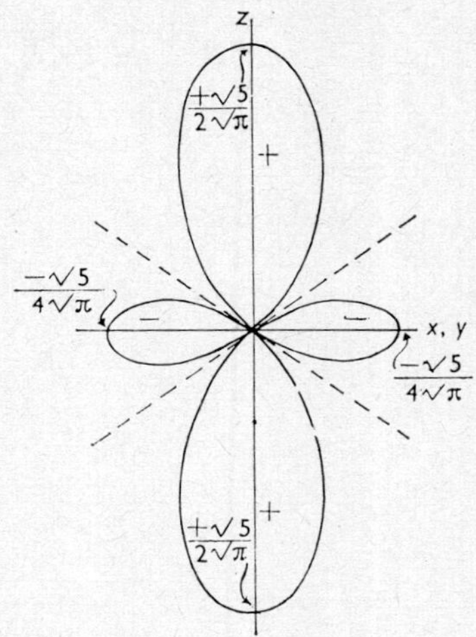

FIG. 7. Polar plot of $A(d_{z^2})$ in the xy plane

cylindrically symmetrical about the z-axis. It has maximum positive values of $\frac{1}{4}\sqrt{(5/\pi)}\,(3-1)$, *ie* $\frac{1}{2}\sqrt{(5/\pi)}$ at $\theta = 0$ and $180°$ and a maximum negative value of $-\frac{1}{4}\sqrt{(5/\pi)}$ at $\theta = 90°$, *ie* in the xy plane. The nodes occur at $(3\cos^2\theta - 1) = 0$. A section through the polar diagram of $A(d_{z^2})$ in the xz plane is shown in *Fig. 7*, the three-dimensional surface being obtained from this by rotation about the z-axis.

The angular dependence functions of the other four d orbitals differ from that of d_{z^2} but are identical in shape with each other and differ only in orientation. Each consists of four lobes alternately positive and negative, as indicated in *Fig. 5*. These shapes are most conveniently verified by considering the Cartesian designation of the orbitals. Thus $A(d_{x^2-y^2})$ has the form of the function (x^2-y^2); it is concentrated near the xy plane, has maximum positive values in the x and $-x$ directions (where y is zero) and maximum negative values in the y and $-y$ directions (where x is zero).

The angular dependence functions of the seven f orbitals follow the same principles. They are illustrated in *Fig. 5* and given explicitly in Table 4.

It will be noticed that the highest power occurring in the θ polynomials defining the angular dependence functions is l; for an s orbital A is a constant, *ie* it varies as z^0 or $\cos^0\theta$; for a p orbital, A depends on z^1 or $\cos^1\theta$ (or on x or y); for a d orbital the dependence is on z^2 or $\cos^2\theta$ or other quadratic expressions. One might think there should be six of these expressions, *viz.* x^2, y^2, z^2, xy, yz and zx, but the first three are not independent, being connected by the relation $x^2+y^2+z^2 = r^2$ (p. 3). There are thus only five independent functions for $l = 2$. Although d_{z^2} has a different shape from the other d orbitals, this does not confer special significance on the z direction since such an orbital can be obtained in any direction simply by taking various linear combinations of d_{x^2}, d_{y^2} and d_{z^2}. Similar arguments apply to the f orbitals; there are 10 possible ternary products of the Cartesian coordinates, but also three equations, leaving only seven independent choices.

The angular dependence function for s orbitals is everywhere of the same sign. For a p orbital it changes sign at the origin as x, y or z goes to $-x$, $-y$ or $-z$, *ie* there is one nodal plane. For d orbitals there are two nodal planes and for f orbitals three. In other words, the total number of nodal surfaces in the angular function is l. We saw on p. 6 that the number of spherical nodes in the radial function was $(n-l-1)$, so the total number of nodes in the orbital, ψ, is $(n-1)$.

Electron spin

The three quantum numbers arose from the necessity of having physically significant solutions of the time-independent Schrödinger equation. If this were all, the 1s orbital of a hydrogen atom would represent a spherically symmetrical, non-degenerate ground state. However, when a beam of atoms is passed through an inhomogeneous magnetic field (the Stern–Gerlach experiment) the beam splits into two equal components. The older quantum theory attempted to explain this, and also the fine structure of the emission spectra of other atoms, by assuming that the electron has a magnetic moment due to spin about its own axis. This *ad hoc* assumption is unnecessary in wave-mechanics, and Dirac has shown that, when the full time-dependent Schrödinger equation is used, and when this is made relativistically satisfactory, then a fourth quantum number, m_s, emerges, which is used in specifying the symmetry or antisymmetry of the wave-function; this quantum number can take one of two values only, sometimes represented by $+\frac{1}{2}$ and $-\frac{1}{2}$. Dirac's system has not been extended to atoms heavier than hydrogen but an alternative, non-relativistic interpretation has been developed for these cases by Pauli (*see* ref. 4, ch. 8).

Although it is not possible to conceive of 'spin' in a wave theory, the term is convenient jargon and is retained in wave-mechanics in the same way that orbitals are used to replace the orbits of older quantum theories. It is also worth pondering on the fact that, if the world were Newtonian, there would be no covalent chemical bonds because these involve spin functions and there is no need for 'spin' in a non-relativistic quantum theory.

Orthogonality

Orthogonality is essentially a simple concept. Two non-identical functions ψ_a and ψ_b are said to be orthogonal if the integral of their product over all space is zero: $\int_{\text{all space}} \psi_a \psi_b \, dv = 0$. If a = b then the integral becomes $\int_{\text{all space}} \psi_a^2 \, dv$ which equals unity if the functions are also normalized. Such functions are called orthonormal functions. It is always possible to arrange that the wave-functions obtained as solutions to the Schrödinger equation are orthogonal to each other and it is important that they should be made so. In fact, precise solutions of the wave-equation are necessarily orthogonal but approximate ones often are not.

The importance of orthogonality is that any arbitrary function can be expressed as a series of orthonormal functions provided certain conditions are satisfied.[5] A familiar example is the Fourier series method which uses sine and cosine functions in x-ray diffraction analysis. The process is also useful in the variation method for finding molecular orbitals and in hybridization theory.

3. Many-Electron Atoms

In the preceding sections we have examined the solutions of the Schrödinger equation for hydrogen-like atoms and discussed explicit expressions for the radial and angular parts of the wave-function. Difficulties are at once encountered when attempts are made to extend this precise treatment to atoms with more than one electron because the potential function, V, in equation (1) on p. 1, now contains terms involving the distances between the various electrons themselves. In other words, the motion of each electron depends not only on its own position with respect to the nucleus but also on the position of all other electrons. As the potential, V, is no longer that of a central force field, the variables r, θ and ϕ cannot be separated and exact analytical solutions are impossible.

Self-consistent field approximation

The difficulty can be resolved by making certain approximations.[6] One electron is selected and imagined to move in the field of the nucleus plus the averaged field of all the other electrons. The wave-functions for this electron can then be defined by a wave-equation which involves only its own coordinates and, as the averaged field of all the other electrons can be made spherically symmetrical, the selected electron can be considered as moving in a central force field.* This means that the potential, $V(r)$, for this electron is independent of θ and ϕ; hence the differential equations are of the same type as for hydrogen and solutions are of the form

$$\psi_{n,l,m}(r,\theta,\phi) = R'_{n,l}(r)\, A_{l,m}(\theta,\phi)$$

In these solutions the radial functions, R', will differ from those for hydrogen orbitals since $V(r)$ is different in the two instances, but the angular parts, A, are precisely the same for both, because the only difference in the equation concerns the variable r which does not occur in the angular function. This is a most important result, for it is this separability that enables us to speak of s, p, d, f orbitals for many-electron atoms and to use the functions in Table 4 (p. 13) and the diagrams in *Fig. 5* to describe them.

It is not possible to write down the averaged field of all the other electrons unless their individual wave-functions are already known. Unfortunately they are not. However, the problem can be resolved

* The field, however, is no longer Coulombic. For example, for uranium the potential would vary continuously in some way from $-e^2/(4\pi\epsilon_0 r)$ at infinity to something like $-92e^2/(4\pi\epsilon_0 r)$ near the nucleus.

by Hartree's self-consistent field method which is to guess plausible wave-functions for all electrons except the one chosen, to calculate their collective potential field and average it over all angles to make it spherically symmetrical, and then to solve the wave-equation for the chosen electron to give a 'first-improved wave-function' for this first electron. A second electron is next chosen and the process repeated using the first-improved wave-function for the first electron and the functions guessed for the remaining electrons. This gives a first-improved wave-function for the second electron. The process is repeated until a set of first-improved wave-functions has been obtained for all the electrons. The whole cycle is then repeated using these first-improved wave-functions instead of the guessed ones in order to obtain a set of second-improved wave-functions. These cycles are repeated until successive iteration leaves all the wave-functions unaltered, this being the self-consistent model for the particular atom. The calculations are done on high-speed computing machines and perhaps 30 iterations might be required before an acceptable degree of consistency is obtained.

The main errors in this method are the neglect of instantaneous interactions between electrons and the smoothing out of the field so that it appears to be spherically symmetrical. Refinements which make the orbitals orthogonal and which take into account the possibility of electron exchange improve by a few per cent the calculated atomic energy-levels but even the best possible one-electron wave-functions lead to energies of atoms which are incorrect by about $40Z$ kJ mol^{-1}, where Z is the number of electrons in the atom. The self-consistent field approximation is better for heavy than for light elements because the attractive force between the highly charged nucleus and the individual electrons is much greater than the inter-electron interaction so that errors made in smoothing out the latter have proportionately less effect.

Several important results emerge from the self-consistent field treatment. We have already mentioned the justification it gives for retaining the s, p, d, f classification of orbitals in many-electron atoms. The radial functions, however, differ considerably from the hydrogen-like radial functions (for results, *see* refs 6, 7, 8). Typical results for some alkali-metal cations are shown in *Fig. 8* which plots the composite radial distribution functions normalized to give the correct total number of electrons in the ions. We see that the electron shells are still discernible, though not completely separated. This arises because, in contrast to older theories of atomic structure in which the K shell, for example, consisted solely of the two 1s electrons, the present calculations show that all electrons in the atom contribute something to each shell. The shells are pulled in closer towards the nucleus with increasing atomic number, the radius of the 1s shell, for

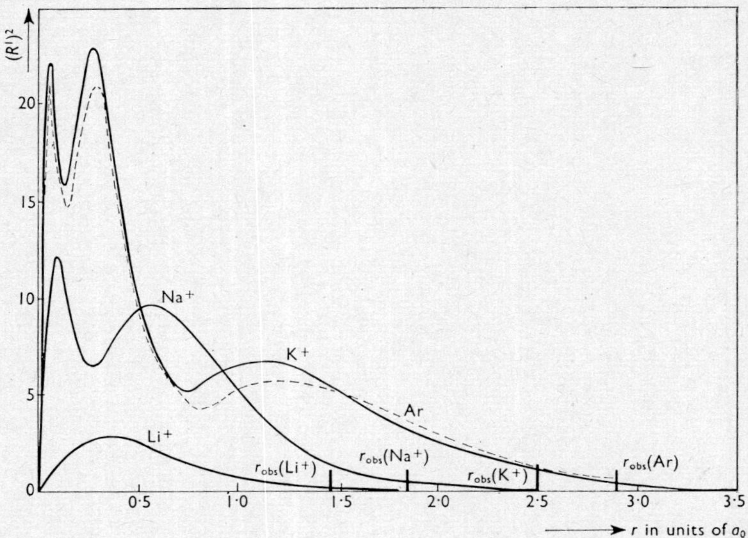

FIG. 8. Radial distribution of electron density for some alkali-metal cations and argon

example, being approximately a_0/Z. *Figure 8* also compares the potassium cation, K^+, with the isoelectronic atom, argon, and shows the contracting effect of the unit positive charge on the ion. This effect is even more noticeable if $4\pi r^2(R')^2$ is plotted instead of $(R')^2$. The conventional empirical ionic radii and the atomic radius of argon are included for comparison.

Electron distributions calculated by the self-consistent field method are obtained in tabular form rather than as analytic functions. This is frequently inconvenient and to overcome this difficulty Slater has devised some approximate wave-functions which can be used for general work and which often reproduce the Hartree distributions quite accurately in cases where these are known. A fuller discussion of the derivation, uses and limitations of Slater functions is given in Appendix 2 (p. 44).

Electron penetration and orbital energies

For a hydrogen-like atom the energy of the electron depends only on the principal quantum number of the orbital it occupies (p. 5). For all other atoms the energy levels (eigenvalues) characteristic of each orbital (eigenfunction) depend not only on n but also on l; that is, the degeneracy of the s, p, d and f energy levels in hydrogen is removed when more than one electron is present in the atom. The

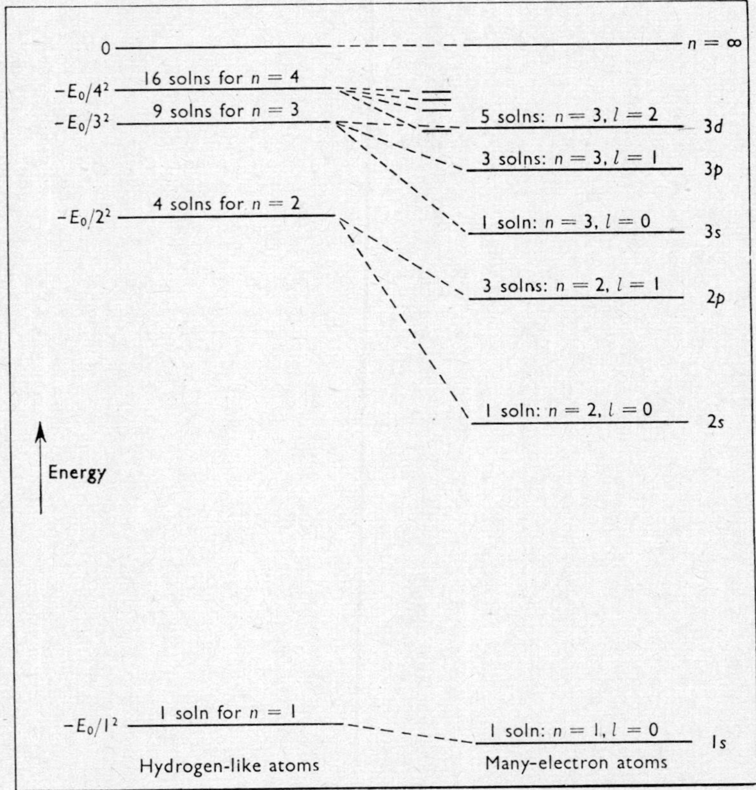

FIG. 9. Effect of electron penetration on energy levels

effect is shown schematically in *Fig. 9* for the typical case of an alkali metal. The order of increasing energy of the occupied orbitals for the first 20 elements follows the sequence $(n+l)$, and if two orbitals have the same value for $(n+l)$ then the one with lower n lies deepest. This implies that the energy is still determined principally by n but is modified by the value of the azimuthal quantum number, l. As we approach higher atomic numbers the sequence of orbital energies tends to rearrange so that it follows the order of the principal quantum number, but for each value of n, the order of decreasing stability is still always s, p, d, f. A similar re-ordering is observed if one increases the nuclear charge without altering the number of electrons (*see* p. 37).

These observations can be understood in terms of the interpenetration of the various orbitals, which results in differences in the

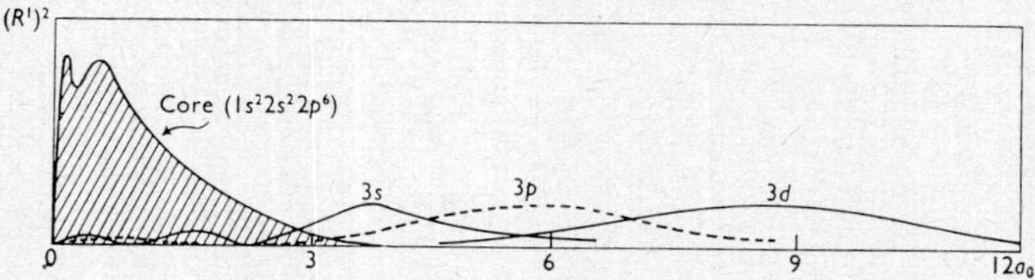

FIG. 10. Radial distribution functions for sodium orbitals

efficiency with which the outer electrons are shielded from the nuclear charge by the inner electrons. The effect is illustrated in *Fig. 10* for the case of the sodium atom. The radial probability function for the 10 electrons in the core ($1s^22s^22p^6$) is shown by the heavy curve bounding the hatched area. The corresponding probability functions for the 3s, 3p and 3d orbitals are also shown. Comparison of the 3s and 3p radial probabilities with those for the corresponding hydrogen orbitals in *Fig. 4* (p. 8) indicates that, because of their penetration into the core where the effective nuclear charge is larger, the 3s and 3p orbitals are greatly reduced in size, the maximum of the last hump occurring at about $3.5a_0$ and $6a_0$ in sodium compared with $13a_0$ and $12a_0$ for hydrogen. By contrast, the 3d orbital remains well outside the main part of the core and the total nuclear charge of $+11$ is effectively reduced to $+1$ by the 10 electrons in the core; the distribution is hydrogen-like and the maximum occurs at $9a_0$ for both sodium and hydrogen.

The concept of orbital penetration is used extensively in discussions of the Periodic Table of the elements (pp. 31–41), and in discussing the order in which the atomic orbitals are occupied.* However, it is important to bear in mind that a given orbital in an atom does not always have the same energy. This is because the energy of an electron in a given orbital depends both on the number of other electrons in the atom and on their configuration. For example, the energy of the 2s orbital in boron is not the same for the two configurations $1s^22s^22p$ and $1s^22s2p^2$, since the shielding of a 2s electron by another 2s electron is more efficient than its shielding by a 2p electron. We see that the energy belongs to the atom as a whole and that energy levels refer to states of the whole atom. Orbital energy is best defined as the energy change which the atom undergoes when

* When an electron is described by an orbital, ψ, *ie* when the probability of finding it at a particular point in the atom is given by $\psi^2 dv$, then the electron is said to occupy the orbital. The phrase 'a $3p_z$ electron' means 'an electron in a $3p_z$ orbital'. This in turn is convenient jargon for 'an electron described by the wave function $\psi_{3,1,0}$, and characterized by the three quantum numbers $n = 3$, $l = 1$, $m = 0$'. The spin quantum number, m_s, is unspecified.

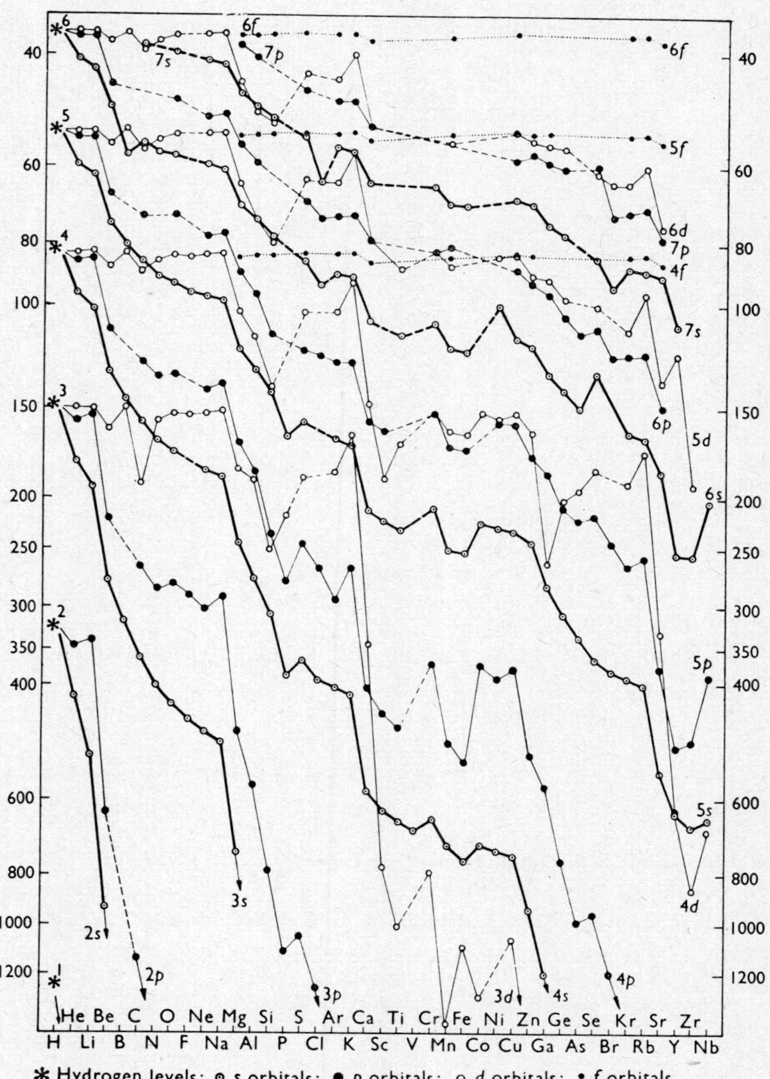

FIG. 11. Orbital energies for the first 41 elements (in kJ mol^{-1})

an electron is removed to infinity from the orbital considered and all other electrons retain their original configuration. These energies are plotted in *Fig. 11* for the first 41 elements and are discussed more fully in the next chapter.

Pauli's exclusion principle

In our discussion so far we have seen that, on the self-consistent field approximation, it is possible to separate out the angular distribution function and then obtain numerical values for the radial distribution functions of the various orbitals. We also saw that the penetration of some orbitals towards the nucleus affected the extent to which they were shielded from the nuclear charge and that this modified their energy. However, we have not yet considered the way in which the extranuclear electrons are distributed among the various possible orbitals, or the total number of electrons that can be described by (contained in) one and the same orbital. It is observed experimentally that no orbital can contain more than two electrons and this led to an enunciation of the Pauli exclusion principle which states, in its modern form, that the total wave-function, including spin states, of systems occurring in nature must be antisymmetric with respect to the interchange of any two electrons in the system. This means that, in a given system, no two electrons can have the same values for all four quantum numbers. As three quantum numbers, n, l, and m, are needed to specify an orbital, it follows that electrons in a given orbital must have different values of the fourth quantum number, m_s. Only two values of the spin quantum number are possible, so each orbital can contain at most two electrons and these must have opposed spins. The detailed application of this principle to the elucidation of the structure of the Periodic Table of the elements is deferred until p. 31.

Resultant quantum numbers

A convenient pictorial method of enumerating electron configurations is to represent the orbitals by groups of square boxes and to indicate whether they are occupied by means of arrows or diagonal lines which also specify the spin quantum number. The convention is that ↓ and ╲ refer to $m_s = -\frac{1}{2}$, whereas ↑ and ╱ refer to $m_s = +\frac{1}{2}$. For example, the ground configuration of the boron atom, $1s^2 2s^2 2p$, can be written in the following equivalent ways:

1s	2s	2p		
↓↑	↓↑	↓		

1s	2s	2p		
⊠	⊠	╲		

For a system with two 2p electrons (*eg* carbon) there are several possibilities; the second p electron could occupy the same orbital as the first p electron, in which case the spins must be opposed, or it could occupy either of the two vacant 2p orbitals, in which case the spins could be either parallel or antiparallel. These states correspond to different energies but, before we discuss their relative stabilities, we must extend our nomenclature to enable us to specify precisely the states being considered.

The resultant azimuthal quantum number, L, of a group of electrons is defined as the vector sum of their individual l quantum numbers, *ie* $L = \Sigma \vec{l}$. Thus, for two p electrons, L can be 2, 1 or 0 as shown below:

$l = 1$, $l = 1$, $L = 2$ $l = 1$, $l = 1$, $L = 1$ $l = 1$, $l = 1$, $L = 0$

It is clear that the vector addition of $\vec{l}+\vec{l}$ can be done by the algebraic addition of the z-component, m, of one to the l value of the other; *ie* $1+1$, $1+0$ and $1-1$. The resultant orbital angular momentum is given by $\sqrt{L(L+1)}(h/2\pi)$ and the resultant states are named by analogy with the one-electron scheme:

$$L = 0, 1, 2, 3, 4, 5, 6, 7, \ldots\ldots$$

$$\text{State} = \text{S, P, D, F, G, H, I, K}, \ldots\ldots$$

The symbol J is omitted to avoid confusion with the resultant inner quantum number, J (*see* below).

The resultant spin quantum number, S, of a group of electrons is the algebraic sum of their individual spin quantum numbers, *ie* $S = \Sigma m_s$, and the z-component of the resultant spin angular momentum is given by $M_{\text{spin}} = \sqrt{S(S+1)}(h/2\pi)$. Thus, for two electrons with parallel spins S is $+1$ or -1, and with opposed spins S is zero. (Note that the symbol S is used for the resultant spin whereas S is used for the state in which $L = 0$.)

The inner quantum number, j, of an electron is defined as the sum of the azimuthal and spin quantum numbers, *ie* $j = l \pm \frac{1}{2}$. The total angular momentum of the electron is then given by the expression $M_{\text{total}} = \sqrt{j(j+1)}(h/2\pi)$. The value of the resultant inner quantum number, J, of a group of electrons will clearly depend on the actual type of coupling which occurs within the atom. One possibility, which is frequently observed in elements of fairly low atomic weight, is Russell–Saunders coupling; here the interaction between the individual orbital momentum vectors themselves to give a resultant L, and the interaction between the individual spin vectors to give a resultant S, are greater than the spin–orbital coupling within each individual electron. The two resultant quantum numbers then couple vectorially to give the resultant inner quantum number $J = \vec{L}+\vec{S}$. For example, for two unpaired p electrons with parallel spins $L = (1+0) = 1$, $S = (\frac{1}{2}+\frac{1}{2}) = 1$ and $J = 2$, 1 or 0. The number of different J-values is called the multiplicity of the state and this equals $(2S+1)$ provided S is not greater than L; if S is greater than L the multiplicity is $(2L+1)$. When Russell–Saunders coupling predominates, the electronic state of an atom can be codified in a 'term symbol', $^{(2S+1)}L_J$, in which the multiplicity and inner quantum number are given as numbers and the resultant azimuthal quantum number, L, is expressed as S, P, D, F, *etc.* Thus the term symbol for the boron configuration shown on p. 24 is $^2P_{1/2}$ and the term symbols for the states arising from the two unpaired p electrons discussed in the present paragraph, which are typical of, for example, the carbon atom, are $^3P_2$, $^3P_1$ and $^3P_0$. By contrast, the term symbol for a carbon atom in the state represented by the left-hand diagram on p. 25 is $^1D_2$, since the resultant L is 2 (*ie* D) and both electrons have $l = +1$, so that their spins must be opposed ($S = 0$).

Another type of coupling between electrons is *jj* coupling. Here the spin–orbital coupling $(l+m_s)$ within each electron to give an inner quantum number j is stronger than the coupling together of all the orbital angular-momentum vectors of the group to give a resultant L, and the coupling of all the spins to give a resultant S. In *jj* coupling the resultant inner quantum number, J, is obtained by summing vectorially all the individual j values. Russell–Saunders and *jj* coupling are limiting cases and intermediate degrees of interaction are possible. For example, in Group IV there is a gradual trend from Russell–Saunders to *jj* coupling with increase in atomic number from carbon to lead.

Hund's rules

When Russell–Saunders coupling predominates, the relative energies of the various possible states (terms) can be correlated by means of Hund's rules. These state that,

(1) for a given electron configuration (*ie* for a given set of n and l quantum numbers) the term with the highest multiplicity (highest S value) is the most stable;

(2) for terms having the same multiplicity, the one with the highest value of L is the most stable;

(3) for a given multiplicity and L value, the term with the lowest J value is the most stable if the shell is less than half-filled; otherwise the term with the highest value of J lies lowest.

Table 5. Some energy levels of carbon

Configuration	*Symbol*	*Carbon orbitals* 2s	2p: +1 0 −1	3s	*Energy/(kJ mol*$^{-1}$*)*
$2s^22p^2$	$^3P_0$	⊠	⧅⧅□		0.0 (ground state)
$2s^22p^2$	$^3P_2$	⊠	⧄⧄□		0.523
$2s^22p^2$	$^1D_2$	⊠	⊠□□		122.0
$2s^22p^2$	$^1S_0$	⊠	□⊠□		259.1
$2s2p^3$	$^5S_2$	⧅	⧅⧅⧅		403.8
$2s^22p3s$	$^3P_0$	⊠	⧅□□	⧅	722.2

An example of these rules is furnished by the energy levels of the carbon atom, some of which are summarized in Table 5. For the configuration $2s^22p^2$, the triplet state with parallel spins lies lower than the singlet state in which the spins are paired (rule 1). Of the two singlet states the D state ($L = 2$) is lower than the S state ($L = 0$) (rule 2); and in the triplet state the term with $J = 0$ is more stable than the term with $J = 2$ (rule 3). In seeking to explain these rules we must penetrate rather deeply into the nature of the forces between two electrons. There are several types of interaction and it is important for many problems in chemistry to know, at least approximately, the magnitude of the energy associated with each.

Let us consider first the electrostatic repulsion between two electrons. If two electrons occupy the same orbital they are necessarily fairly close together and their mutual electrostatic repulsion will raise the potential energy of the system. For example, the repulsion energy of two electrons in a 1s orbital is 1.25 MJ mol^{-1} for the hydride ion H^-, and 2.88 MJ mol^{-1} for helium. If there were another orbital lying close to the 1s in energy but with a different spatial distribution, then it would be energetically favourable to uncouple the spins and place one electron in each orbital. Such a situation arises, for example, in carbon, which has the configuration $1s^22s^22p^2$. The three 2p orbitals are degenerate (have the same energy) so the most favourable distribution is to place one electron in $2p_x$ and one in $2p_z$; this reduces electrostatic repulsion to a minimum since there is zero probability of finding the p_z electron in the x-direction for which p_x has its maximum probability, and zero probability of finding p_x along the *z*-axis. The minimum energy required to place both electrons with paired spins in the same space orbital is about 122 kJ mol^{-1}, as may be seen by comparing the energies of the $^3P_0$ and $^1D_2$ states in Table 5. This energy, which represents the difference in electrostatic repulsion in the two states, is much less than the electrostatic repulsion of two electrons in the 1s orbital of helium which is unusually large because the 1s orbital does not extend very far out from the nucleus and the two electrons are confined together in a very small volume.

The second type of force between two electrons is rather more subtle and arises from the Pauli exclusion principle, as a result of which two electrons with the same spin have a low probability of being near one another.[9] Pauli forces are not electrostatic or electrodynamic in origin; they act by controlling the movement of the electrons which in turn affects the energy. This can be seen by considering the 1s2s state of helium which can be singly occupied in either of two ways:

The wave-function for the state will be based on the product

$$\psi(a,b) = \psi_{1s}(a)\,\psi_{2s}(b)$$

where the two electrons are denoted by 'a' and 'b'. This product wave-function by itself is unsatisfactory as a solution to the time-independent Schrödinger equation of the system because its symmetry is incorrect with respect to interchange of the electrons. Satisfactory solutions are obtained by linear combinations:

$$\psi_{\text{sym}}(a,b) = \frac{1}{\sqrt{2}}\,[\psi_{1s}(a)\,\psi_{2s}(b) + \psi_{1s}(b)\,\psi_{2s}(a)]$$

$$\psi_{\text{antisym}}(a,b) = \frac{1}{\sqrt{2}}\,[\psi_{1s}(a)\,\psi_{2s}(b) - \psi_{1s}(b)\,\psi_{2s}(a)]$$

According to Pauli's exclusion principle, the combination of these space functions with the corresponding spin functions must have overall antisymmetry, *ie* the spin function for ψ_{sym} must be antisymmetrical (spins opposed) and the spin function for ψ_{antisym} must be symmetrical (spins parallel). Graphical plots of the solutions[9] show that ψ_{sym} favours charge distributions in which the electrons are close together, whereas ψ_{antisym} favours charge distributions in which the electrons are well separated. In other words, Pauli forces keep electrons with parallel spins apart and thus reduce their electrostatic repulsion. The difference in energy of the two states in the case of helium is 77.0 kJ mol^{-1}. The effect is quite general and immediately explains Hund's first rule that the state having the highest multiplicity has the lowest energy, since the maximum multiplicity is obtained by aligning the electron spins parallel.

Hund's second rule states that, among terms of the same multiplicity, the one with the highest resultant orbital angular momentum has the lowest energy. This can perhaps best be understood by observing that high values of the resultant orbital angular momentum arise when the angular momenta of the individual electrons reinforce each other. In terms of a mechanical model this means that the electrons tend to move in the same direction; they therefore collide less frequently and are, on average, further apart than if their angular momenta were opposed. The electrostatic repulsion between the electrons is therefore minimized with consequent stabilization of the system. The energy differences involved are similar to those resulting from the operation of Pauli forces in aligning the electron spins. For example, we see from Table 5 that, of the two singlet states of carbon, the $^1D_2$ state for which $L = 2$ lies 137 kJ mol^{-1} below the $^1S_0$ state for which $L = 0$, whereas the difference in energy between the lowest singlet and the triplet state is 122 kJ mol^{-1}.

Hund's third rule deals with states having different J values, and the energy differences here are much smaller, usually less than 1–2 kJ mol^{-1}. Thus, in the ground-state triplet of carbon, the $^3P_1$ term lies 0.197 kJ mol^{-1} above the lowest term $^3P_0$, and the $^3P_2$ term lies 0.523 kJ mol^{-1} above the triplet $^3P_0$ (*see* Table 5). The different energies arise from the magnetic coupling of the resultant spin and orbital angular momentum. Both the spin and the orbital motion are associated with magnetic moments, and the energy of the system depends on the relative orientation of these moments. For singly-occupied orbitals, the state of lowest energy is when the spin magnetic moment is antiparallel to the orbital magnetic moment ($J = L - S$) and the state of highest energy is the one in which the two moments are parallel ($J = L + S$) as required by Hund's rule.

We see that the observed sequence of electronic energy-levels in an atom results from the interplay of a variety of forces. Orbital penetration determines the shielding efficiency and hence influences the electrostatic attraction between electrons and the nucleus. Electrostatic repulsions between the electrons themselves depend on the particular space orbitals they occupy and, more subtly, on the correlations resulting from the Pauli forces which in turn depend on the alignment of electron spins. Electrodynamic forces depend on the magnitude of the total orbital angular momentum, and the fine structure derives from magnetic interactions stemming from spin–

orbital coupling. Because of the complexity of the problem it is not always possible to predict from first principles the expected sequence of energy levels for each atom, but the observed sequence can almost always be interpreted in terms of the preceding concepts. In this sense the interpretation of the form of the Periodic Table and the detailed analysis of atomic energy-levels as revealed by x-ray and emission spectra, are a joint triumph of the chemist and the physicist.

4. The Periodic Classification

On the basis of Pauli's exclusion principle, the number of orbitals in each quantum shell can be predicted. When $n = 1$, both l and m must be zero (*see* appendix 1, p. 42) so that only one orbital (1s) appears in the first shell and this can contain two electrons with opposed spins. When $n = 2$, l can be 0 or 1 (s or p); for the 2s orbital m can only be zero as before but for 2p, m can take on the values ± 1 or 0 giving three possibilities p_x, p_y and p_z. Each can hold two electrons so the shell will be complete when there are two 2s and six 2p electrons in the atom. For $n = 3$, l can have the values 0, 1 or 2. Taking the m quantum number into account we see that there will be one 3s and three 3p orbitals as before, and also five 3d orbitals ($m = \pm 2, \pm 1$ or 0). This means that there is room for 18 electrons (*ie* $2+6+10$) in the third quantum shell, and this expands by a further 14 to 32 in the fourth shell where f orbitals are possible ($m = \pm 3, \pm 2, \pm 1$ or 0). If the energies of these various orbitals followed the sequence calculated for the hydrogen atom, the derivation of a periodic classification for the elements would follow immediately. However, we have seen that penetration effects (p. 21) modify the energy sequence considerably in the third and later quantum shells so that the 4s orbital is occupied before electrons start filling the 3d orbitals, and entry into the 4f orbitals is delayed until after the 5s, 5p and 6s orbitals have been filled.

The existence of these various groups of orbitals is strikingly demonstrated by a graph of the ionization energies of the elements (*see Fig. 12*). The ionization energy is the energy required to remove one electron from an isolated atom to infinity and is a measure of how tightly the particular electron is bound within the atom. A detailed knowledge of the electron configurations, energy levels and ionization energies of the elements is fundamental to an understanding of their chemical properties and a discussion of individual elements now follows.*

Hydrogen, helium and the first short period

The energy-level diagram for the hydrogen atom can be calculated accurately from the precise solution of the Schrödinger equation (p. 3) and the results agree with experiment. The ground configuration is 1s and this lies 983.2 kJ mol^{-1} below the next level, 2s (or 2p).

* Energy levels taken from NBS Circular 467[10] and from Landolt–Börnstein's Tables.[8] (1 kJ mol^{-1} = 83.62 cm^{-1}; 1 cm^{-1} = 11.96 J mol^{-1} = 1.2394×10^{-4} eV $atom^{-1}$.)

1s	1s	1s 2s	1s 2s
↑	↑↓	↑↓ ↑	↑↓ ↑↓
Hydrogen ($^2S_{1/2}$)	Helium ($^1S_0$)	Lithium ($^2S_{1/2}$)	Beryllium ($^1S_0$)

In building up the electronic configuration of helium we start first with one electron in the 1s ground configuration of He^+. To place a second electron in the same orbital involves an interelectronic repulsion energy of 2880 kJ mol^{-1}; this is less than the energy of the next available level in He^+ which lies 3940 kJ mol^{-1} above the ground state, so the configuration of the helium atom is $1s^2$. The first excited state, 1s2s ($^3S_1$), has two unpaired electrons but this lies 1912 kJ mol^{-1} above the ground state and so can not be used for covalent bonding since bond energies in excess of 950 kJ/bond would be needed to repay the excitation energy. (The highest known covalent single-bond energy is 565 kJ mol^{-1} for the H–F bond.) Likewise, as shown in *Fig. 12*, the high ionization energy of 2372 kJ mol^{-1} for helium precludes the possibility of cationic compounds for this element. However, compounds containing He^- are conceivable.

The next element, lithium, has three electrons. Since Pauli's exclusion principle limits the first quantum shell to two electrons, the third electron must be placed in the next available orbital, giving a configuration of $1s^22s$. Excitation to the $1s^22p$ state requires 175 kJ mol^{-1}; this involves an increase in l but not in n and is much less than is required to excite hydrogen which involves an increase in the principal quantum number. The excitation energy illustrates the extent to which the 2s orbital is stabilized by penetrating through the core of 1s electrons to regions near the nucleus where it is less efficiently shielded than the 2p orbital. *Figure 12* shows that the 2s electron in lithium is much less tightly bound than the 1s electrons in hydrogen and helium. Ionic compounds of the univalent cation Li^+ are therefore to be expected.

Beryllium parallels the situation in helium. The placing of a second electron in the 2s orbital involves strong interelectronic repulsion but this is energetically more favourable than the configuration $1s^22s2p$ ($^3P_0$) which lies 263 kJ mol^{-1} above the ground-state configuration, $1s^22s^2$. We see again that the energy required for excitation from 2s to 2p is much less than for 1s to 2s (1912 kJ mol^{-1} for helium); were it not for this, beryllium, having no unpaired electrons in the ground state, would behave chemically like the inert gas helium. As it is, the promotion energy is less than the energy released in forming bonds with the two unpaired electrons, and beryllium is divalent.

In boron, the 2p orbitals begin to fill, and the electronic configura-

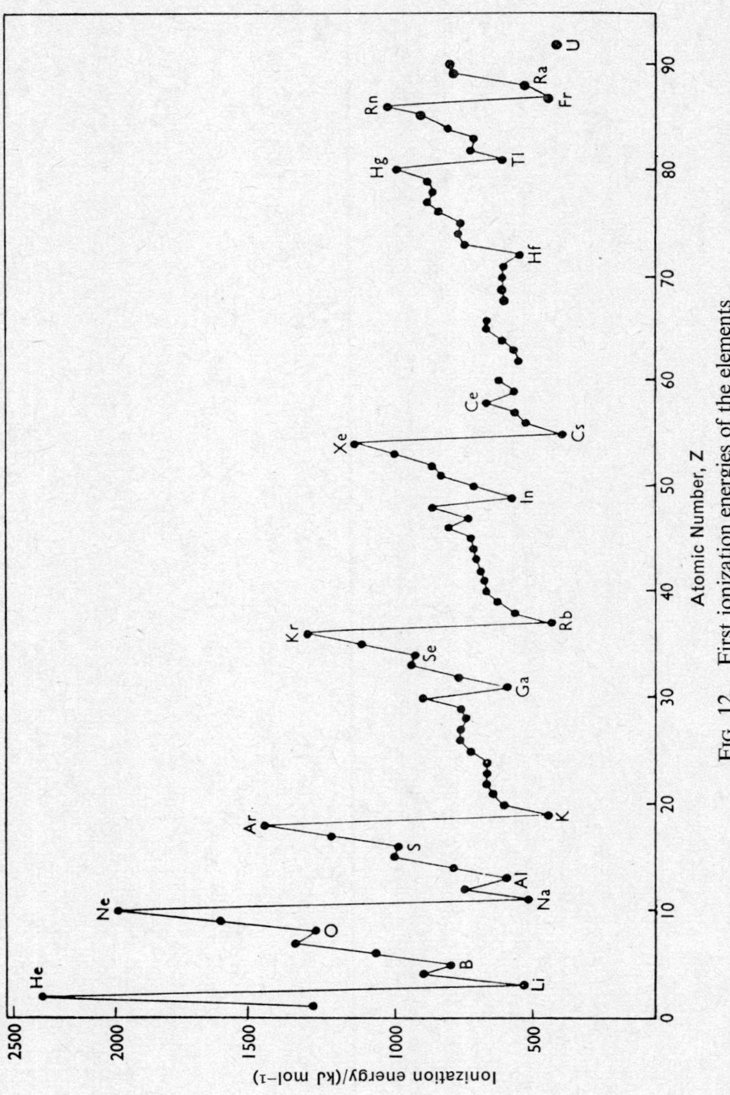

FIG. 12. First ionization energies of the elements

tion of this element and of carbon and nitrogen are as shown, the 1s orbital being omitted in each case for convenience.

2s 2p	2s 2p	2s 2p
Boron ($^2P_{1/2}$)	Carbon ($^3P_0$)	Nitrogen ($^4S_{3/2}$)

The diagrams show that, in accordance with Hund's rules, the 2p orbitals are filled singly and that the electron spins are aligned parallel to each other and against the orbital momentum (though it needs only 0.193 kJ mol^{-1} to align the spin and orbital magnetic moments parallel in boron). As with beryllium, only a moderate amount of energy is required to excite one of the 2s electrons in boron or carbon into a 2p orbital.

$$\text{Boron} \quad 2s^22p\ (^2P_{1/2}) \longrightarrow 2s2p^2\ (^4P_{1/2}) = 345 \text{ kJ mol}^{-1}$$

$$\text{Carbon} \quad 2s^22p^2\ (^3P_0) \longrightarrow 2s2p^3\ (^5S_2) = 404 \text{ kJ mol}^{-1}$$

This increases the number of unpaired electrons and permits these elements to be tervalent and quadrivalent, respectively. There is also a promotion energy from the excited state to the valency state, in which the electrons, instead of having their spins aligned parallel, are uncorrelated. In the case of carbon this involves a further 234 kJ mol^{-1} so that the valency state is some 636 kJ mol^{-1} above the ground state. As the C–H bond energy is approx. 414 kJ mol^{-1}, the formation of the two extra bonds stabilizes the system by about 192 kJ mol^{-1}. It is clear that if the promotion energy from $^3P_0$ to $^5S_2$ in carbon had been somewhat higher, or the C–H bond energy slightly lower, then CH_2 would have been appreciably more stable than CH_4 and carbon would have been predominantly divalent.

As regards nitrogen, the three 2p orbitals are already singly occupied, and the promotion, besides needing more energy, leaves the number of unpaired electrons unchanged.

$$\text{Nitrogen} \quad 2s^22p^3\ (^4S_{3/2}) \longrightarrow 2s2p^4\ (^4P_{5/2}) = 1054 \text{ kJ mol}^{-1}$$

Excitation to $2s2p^33s$ ($^6S_{5/2}$) which would enable nitrogen to be quinquivalent clearly requires considerably more energy, and transitions involving this level have not been observed spectroscopically.

In oxygen and fluorine, the process of filling the 2p orbitals continues and they are completely filled in neon.

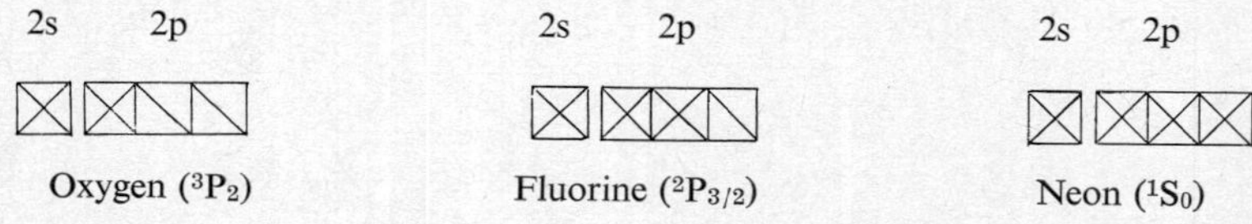

Oxygen ($^3P_2$) Fluorine ($^2P_{3/2}$) Neon ($^1S_0$)

This process involves placing a second electron successively in each of the three 2p orbitals. These electrons are repelled by the electrons already in the orbitals, *ie* they are less tightly bound than they would otherwise be. This is reflected in the discontinuous drop in ionization potential which occurs between nitrogen and oxygen, as shown in *Fig. 12*.

The ground state of oxygen, $^3P_2$, is inverted, as expected from Hund's third rule, and it requires 2.72 kJ mol^{-1} to alter the spin–orbital coupling to give $^3P_0$. To pair all the spins ($^1D_2$) increases the energy by 190 kJ mol^{-1}, whereas the uncoupling and promotion of one of the paired 2p electrons to give four unpaired electrons, $2s^22p^33s$ ($^5S_2$), involves 883 kJ mol^{-1}; this is evidently too much to allow oxygen to be normally quadrivalent. Promotion in fluorine to give a state with three unpaired electrons, $2s^22p^43s$ ($^4P_{5/2}$), is even less probable, 1226 kJ mol^{-1}, and with neon the energy difference between the $2p^6$ ground state and the $2p^53s$ level with two unpaired electrons is 1602 kJ mol^{-1}. The reason for considering these figures in detail is to acquire a background for the discussion of valency problems. Thus, from the first 10 elements we have learned that excitations of less than 400 kJ mol^{-1} appear feasible in chemical reactions, whereas excitations much above 800 kJ mol^{-1} are not observed.

Second short period

The 3s orbital is the next most stable after the neon core has been filled, and in sodium and magnesium this is occupied by the 11th and 12th electrons. The first excited state in magnesium is 3s3p ($^3P_0$) and this lies 262 kJ mol^{-1} above the $3s^2$ ($^1S_0$) ground state; this is almost identical with the corresponding 2s → 2p promotion energy for beryllium, but the process is not so important in the chemistry of magnesium since the lower ionization energies of the heavier element permit the formation of ionic rather than covalent compounds.

The electron configurations of the elements from aluminium to argon show that the 3p orbitals fill in the same way as the 2p orbitals, at first singly, then doubly, as described by Hund's rules.

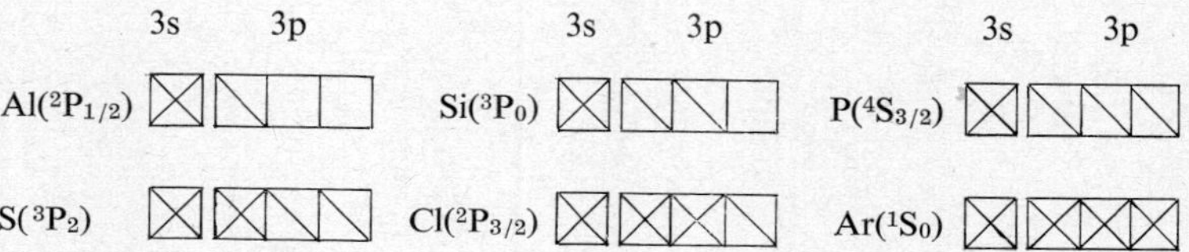

Aluminium differs from boron in that its first excited state (at 303 kJ mol^{-1}) is $3s^24s$ ($^2S_{1/2}$), and 347 kJ mol^{-1} are required to reach the quartet state $3s3p^2$ ($^4P_{1/2}$) with three unpaired electrons; this is very similar to the energy required for the analogous excitation in boron which we found to be 345 kJ mol^{-1}. Transitions in silicon, phosphorus and sulphur to states involving four, five and six unpaired electrons have not been observed spectroscopically. Promotion in chlorine to give a state with three unpaired electrons, $3s^23p^44s$ ($^4P_{5/2}$), demands 862 kJ mol^{-1}. Although high, this evidently does not preclude the formation of tervalent chlorine in compounds like ClF_3; the excitation energy is 364 kJ mol^{-1} less than the analogous excitation in fluorine, which is never tervalent.

First and second long periods

In argon the 3s and 3p orbitals are completely filled. The 19th electron in the next element, potassium, does not enter a 3d orbital, however, because orbital penetration makes the 4s level more stable than the 3d for elements with atomic number greater than six (carbon). The detailed sequence of energy levels is shown in *Fig. 11* (p. 23), which plots on a logarithmic scale the energy required to remove an electron from the stated orbital. It illustrates the degeneracy of s, p, d and f orbitals in hydrogen, and the increasing effect of orbital penetration in removing this degeneracy in many-electron atoms. As was pointed out earlier (*see* p. 22), the d and f orbitals remain well outside the core electrons for most of the lighter atoms; they are thus almost completely shielded from the increasing nuclear charge and their energies are virtually the same as those of the corresponding orbitals in hydrogen. For example, *Fig. 11* shows that the energy required to remove an electron from a 3d orbital is 146 kJ mol^{-1} for hydrogen and 147 kJ mol^{-1} for sodium. A similar situation obtains for potassium though there is a slight increase in the binding energy of the 3d orbital to 162 kJ mol^{-1}. The situation is quite different with calcium which has the ground-state configuration $3s^23p^64s^2$, because the 4s orbital overlaps the 3d orbital considerably. It follows that in the excited state, $3s^23p^63d4s$, the electron in the 3d orbital is no longer completely outside the core ($3s^23p^64s$) and is therefore less efficiently shielded from the increased nuclear charge; its binding energy increases dramatically to 346 kJ mol^{-1}.

The process continues in the next element, scandium (ground state $3d4s^2$) and the shielding of the 3d orbitals by the two 4s electrons is so inefficient that the 3d level falls not only below the 4p but also below the 4s level itself, and for the first time in the Periodic Table a d orbital is occupied. The five 3d orbitals can accommodate 10 electrons and this accounts for the appearance of the first series of transition elements from scandium to zinc. We see that the occurrence of the transition elements at this particular point in the Table is to some extent fortuitous because, if the penetration stabilization of the 3d levels had proceeded no further than it had done at calcium, then the 4p level would have been the next to fill and a third short period would have resulted. Again, we see from *Fig. 11* (p. 23) that there is a rapid stabilization of the 3d levels after sodium and magnesium and, if this had continued, the transition series would have started after the completion of the 3p shell, *ie* potassium would have been a transition element with the configuration $3s^2 3p^6 3d$.

In scandium and the succeeding elements the shielding of electrons in the 3d and 4s orbitals by each other is a mutual effect and interelectronic repulsion terms also become important in determining the configuration of lowest energy. For this reason all three outer electrons in scandium do not go into 3d orbitals; the configurations $3d^2 4s$, 3d4s4p and $3d^3$ lie 138, 187, and 404 kJ mol^{-1} above the $3d4s^2$ ground state.

In *Fig. 11* it is shown to be easier by 141 kJ mol^{-1} to remove a 4s than a 3d electron from the scandium atom and the configuration of the Sc^+ ion is accordingly 3d4s. The ion has the same number of electrons as the neutral calcium atom, $4s^2$, and the change in configuration reflects the influence of the larger nuclear charge on the inefficiently shielded 3d orbital. Similarly the Sc^{2+} ion, which is isoelectronic with potassium, has the configuration 3d. In Table 6 more extensive data are summarized for these and neighbouring isoelectronic series, the effect of the increasing effective nuclear charge in each vertical column being to stabilize the 3d levels relative to the 4s.*

After scandium the 3d orbitals fill progressively as shown in Table 7 and this is reflected in the ionization energies in *Fig. 12* (p. 33). In each configuration the spins are as far as possible aligned parallel and the 3d energy levels lie below the 4s. The filling is not quite regular just before the half-filled and completely-filled sub-shell because of the electronic interactions mentioned above (p. 27). Thus, for chromium, the $d^5 s^1$ ($^7S_3$) configuration is more stable than the $d^4 s^2$ ($^5D_0$) configuration by 93 kJ mol^{-1}.

There are three reasons for this in addition to the obvious one of differing coulomb attraction between the nucleus and a 3d or a 4s electron charge distribution. Firstly, the $d^5 s^1$ configuration places each of the six electrons in a different orthogonal orbital and this

* This does not imply that a d-type orbital is stabilized more than an s-type orbital. A 4s orbital penetrates nearer the nucleus and is stabilized more by the increased effective nuclear charge than is a 4d orbital, and a 3s orbital is bound more firmly than a 3d in any multi-electron atom. However, the binding energy also depends on Z^2 for hydrogen (p. 5) and $(Z^*)^2$ for other atoms (see Appendix 2, p. 44) so that the vertical distance between the levels corresponding to $n = 1, 2, 3, 4 \ldots$ in *Fig. 11*, p. 23 (and in *Fig. 9*, p. 21) becomes increased to such an extent that, despite considerable stabilization by penetration, the 4s level does not fall below the 3d level. Hence ionization stabilizes 3d relative to 4s.

Table 6. Ground-state configurations for isoelectronic series

19		20		21		22		23 (*electrons*)	
K^+	$-4s$								
Ca^+	$-4s$	Ca	$-4s^2$						
Sc^{2+}	$3d-$	Sc^+	$3d4s$	Sc	$3d4s^2$				
Ti^{3+}	$3d-$	Ti^{2+}	$3d^2-$	Ti^+	$3d^24s$	Ti	$3d^24s^2$		
V^{4+}	$3d-$	V^{3+}	$3d^2-$	V^{2+}	$3d^3-$	V^+	$3d^4-$	V	$3d^34s^2$

Table 7. Electron configuration of first-row transition elements

Element	*Sc*	*Ti*	*V*	*Cr*	*Mn*
Configuration	$3d^14s^2$	$3d^24s^2$	$3d^34s^2$	$3d^54s^1$	$3d^54s^2$
Term symbol	$^2D_{3/2}$	$^3F_2$	$^4F_{3/2}$	$^7S_3$	$^6S_{5/2}$
Element	*Fe*	*Co*	*Ni*	*Cu*	*Zn*
Configuration	$3d^64s^2$	$3d^74s^2$	$3d^84s^2$	$3d^{10}4s^1$	$3d^{10}4s^2$
Term symbol	$^5D_4$	$^4F_{9/2}$	$^3F_4$	$^2S_{1/2}$	$^1S_0$

reduces the electrostatic repulsion inherent in the second configuration, which has two electrons in the 4s orbital. Secondly, it minimizes the orbital angular momentum, which is zero instead of $\sqrt{2(2+1)}(h/2\pi)$. Thirdly, and this is probably the most important factor, all six spins are aligned parallel. This means that there is maximum possibility of electron exchange and hence maximum stabilization of the configuration by the exchange energy. In other words, the Pauli forces impose an electron correlation which is favourable to the potential energy of the system (*see* p. 28). The extent of the stabilization due to this third factor can be gauged by comparing the energies of the d^5s^1 configuration in its ground state ($^7S_3$) and its first excited state ($^5S_2$) where the energy required to hold the spin of one electron antiparallel to the spins of the other five is 91 kJ mol^{-1}. This is only 2 kJ mol^{-1} less than the energy of excitation from the ground-state configuration to d^4s^2, in which again one spin is antiparallel to the other five.

With copper the situation is rather different because the ground state $d^{10}s^1$ ($^2S_{1/2}$) and the first excited configuration d^9s^2 ($^2D_{5/2}$) each have only one unpaired electron, so that exchange energies and Pauli forces are not involved. The difference of 134 kJ mol^{-1} between the energies of the two states presumably reflects, among other things: (*a*) the difference in electrostatic attraction between the nucleus and a 3d or a 4s electron charge distribution; (*b*) the difference in electrostatic repulsion between a pair of electrons in a 3d orbital in the first configuration and in a 4s orbital in the second;

(*c*) the difference in orbital angular momentum in the two states—zero and $\sqrt{6}(h/2\pi)$.

The next element, zinc, has the configuration $3d^{10}4s^2$ and this completes the first series of transition elements. The similarity in ionization energies and orbital energies of the various configurations of these elements leads to the possibility of variable valency, and the transition elements are notable for their ability to exist in a variety of oxidation states as successive electrons become involved in bonding, *eg* Mn^{II}, Mn^{III}, Mn^{IV}, Mn^{V}, Mn^{VI} and Mn^{VII}. In this they differ from the main group elements (p orbitals) in which variable valency involves uncoupling a pair of electrons so that oxidation numbers change in steps of two, *eg* SCl_2, SF_4, SF_6.

In view of the ability of copper to form both copper(I) and copper(II) compounds [Cu(0), $d^{10}s^1$; Cu(I), d^{10}; Cu(II), d^9] it becomes of interest to inquire whether zinc can also exist in the tervalent state [Zn(III), d^9]. The third-stage ionization energy to give Zn^{3+}, though high, is not prohibitively so, and promotion of a 3d electron to the 4p level should also be possible.

Reference to *Fig. 11* (p. 23) shows that, after the 3d and 4s subshells have been completed in zinc, the next available orbitals are the 4p. These fill regularly in the next six elements Ga, Ge, As, Se, Br and Kr, and this completes the first long period.

After element 36, krypton, there follows an alkali metal, rubidium, and an alkaline-earth metal, strontium, in which the last added electrons enter the 5s orbital. As the spatial distribution of this 5s orbital overlaps the 4d orbitals, the latter become less effectively shielded and hence are stabilized as shown in *Fig. 11*. A second transition series ensues from yttrium to cadmium. Although analogous to the first transition series there are differences in detail. For example, *Fig. 11* indicates that the stabilization of the 4d orbitals in yttrium has not proceeded to the extent that they lie below the 5s orbital; in agreement with this, ionization of Y(ds^2) gives Y^+(s^2) and not ds as with scandium. In subsequent elements, however, the 4d orbitals lie below the 5s and, indeed, from niobium onwards, the ground state is d^ns^1 rather than d^ms^2. The only exceptions are Pd ($d^{10}s^0$) and Cd ($d^{10}s^2$) at the end of the series. A complete list of ground-state configurations and ionization energies for the elements is given in the Periodic Table on p. 47.

The second long period ends with the regular filling of p orbitals in the six elements In, Sn, Sb, Te, I and Xe. The chemistry of these elements continues the trends observed in their lighter congeners, and the tendency towards electropositive behaviour becomes more noticeable. This arises from the decrease in promotion energies with increasing atomic number in a group, a general trend which follows from the inverse dependence of orbital energy on n^2, where n

is the principal quantum number. This has an interesting consequence in the inert-gas group, because the energy required to uncouple two of the np^6 electrons and promote one of them to the $(n+1)$s level decreases for the heavier gases to values which permit chemical reaction under the right conditions. The promotion energies are summarized in Table 8. Various theories of the bonding in compounds of krypton, xenon and radon have been proposed[11] but the Table illustrates the inherent plausibility of compounds of these elements.

Table 8. Promotion energies to the $np^5(n+1)$s state in inert gases

Element	*Ne*	*Ar*	*Kr*	*Xe*	*Rn*
Energy/(kJ mol^{-1})	1602	1113	958	699	653

Heavy elements

After xenon, the 6s orbital fills in caesium and barium, and the third transition series (5d) starts at the next element, lanthanum. A new complication arises here, however, because the 6s and 5d orbitals overlap the 4f orbitals which up to this point have been well outside the electron charge distribution of the occupied orbitals and have retained an orbital energy very similar to that of hydrogen itself. The shielding of the 4f orbitals by the core electrons is so efficient that, even in caesium with a nuclear charge of 55, the 4f binding energy of 82.8 kJ mol^{-1} is only 0.95 kJ mol^{-1} greater than in hydrogen itself. In barium, however, the 4f binding energy increases to 162 and in lanthanum to 255 kJ mol^{-1}. The process continues with cerium in which the binding energy of the 4f orbital becomes greater than that of both the 6p and the 5d orbitals and the ground-state configuration is accordingly $4f^2 5s^2 5p^6 6s^2$. Thereafter, the 4f orbitals fill singly until the sub-shell is half full at europium, $4f^7 6s^2$. Gadolinium follows with the configuration $4f^7 5d^1 6s^2$, but in the succeeding elements the 4f orbitals are again favoured, terbium being $4f^9 6s^2$ and so on to ytterbium, $4f^{14} 6s^2$, and lutetium, $4f^{14} 5d^1 6s^2$. Details are given in the Periodic Table on p. 47.

The third transition series (5d) then continues but, as expected, the underlying core of 14 4f electrons modifies the detailed order of occupation. Thus, there is no anomaly at the half-filled stage, tungsten having the configuration $5d^4 6s^2$ and rhenium $5d^5 6s^2$. The ground states of the last five elements in the series are Os, $d^6 s^2$; Ir, $d^9 s^0$; Pt, $d^9 s^1$; Au, $d^{10} s^1$; Hg, $d^{10} s^2$. These deviations from regularity in the electron configurations of the transition elements emphasize again the complexity of the interactions which determine the ground state and the consequent proximity in energy of alternative configurations. As a result, the elements in six out of the 10 transition groups of the Periodic Table do not have identical outer electron configurations for the ground state of the isolated atoms, *eg* Ni, $3d^8 4s^2$; Pd, $4d^{10} 5s^0$; Pt, $5d^9 6s^1$.

After the completion of the 6p shell at radon, the 7s orbital is occupied in francium and radium and the fourth transition series (6d) then begins with actinium $6d^1 7s^2$. Unlike the 5d series, however, which was immediately superseded after lanthanum by the 14 4f lanthanide elements, actinium is followed by thorium, $6d^2 7s^2$, and it is only in subsequent elements that the 5f levels become occupied: Pa, $5f^2 6d^1 7s^2$; U, $5f^3 6d^1 7s^2$ *etc.* as listed on p. 47.

In this chapter we have seen that the basic wave-mechanical picture of atomic orbitals (which was developed in chapters 2 and 3) can be used to give a detailed description of the electronic structure of the elements. In addition, it has been stressed that chemical properties depend not only on the ground-state electronic configuration of an element but also on the energy of excitation to alternative states, on the ease of ionization and on whether the d and f orbitals underlying the valency electrons are occupied or not. It is also clear that the directional properties of the angular dependence functions of atomic orbitals are of prime importance in discussing the stereochemistry of covalent compounds, but a detailed discussion of these topics is beyond the scope of this Monograph.

Appendix 1. Solution of the Θ and *R* Equations

Substitution of the function

$$\psi(r,\theta,\phi) = R_{n,l}(r)\ \Theta_{l,m}(\theta)\ \Phi_m(\phi)$$

in the time-independent Schrödinger equation for the hydrogen atom (p. 3) leads to three simpler differential equations each in one variable only. The solution of the Φ equation was discussed on p. 4. Solutions of the Θ and R equations are more complicated and explicit expressions for the functions $\Theta_{l,m}(\theta)$ and $R_{n,l}(r)$ require the definition of two types of function familiar to mathematicians, *viz.* associated Legendre polynomials and associated Laguerre polynomials.

A Legendre polynomial, $P_l(x)$, of degree l in the variable x is obtained by differentiating the expression $(x^2-1)^l$ l times and then dividing the result by $2^l l!$

$$P_l(x) = \frac{1}{2^l\, l!}\frac{\mathrm{d}^l(x^2-1)^l}{\mathrm{d}x^l} \qquad 1$$

By subsidiary definition $P_0(x) = 1$. An associated Legendre polynomial, $P_l^m(x)$, is obtained by differentiating the corresponding Legendre polynomial m times and then multiplying the result by $(1-x^2)^{m/2}$

$$P_l^m(x) = (1-x^2)^{m/2}\frac{\mathrm{d}^m P_l(x)}{\mathrm{d}x^m} \qquad 2$$

In wave-mechanics, associated Legendre functions occur in solutions of the angular part of the wave-equation, $\Theta_{l,m}(\theta)$, in which they take the form $P_l^{|m|}(\cos\theta)$.

The orbital quantum number, l, arises because the power series by which the solution is expressed must be cut off at a finite number of terms for a non-infinite solution, and this can only be achieved by introducing a constant β in the form $\beta = l\,(l+1)$, where l is zero or a positive integer. Having expressed the solution in terms of associated Legendre functions it is clear that l must be integral since a non-integral number of differentiations is meaningless.

Equation 1 shows that the highest power of x occurring in $P_l(x)$ is x^l. Differentiation m times reduces this to x^{l-m}. Hence if $|m|$ is greater than l, then $P_l^{|m|}(x)$ vanishes. This means that $|m|$ must not be greater than l; that is m can run from $-l$ to $+l$ and has $(2l+1)$ values.

A Laguerre polynomial, $L_k(x)$, of degree k in the variable x is obtained by differentiating the function $x^k e^{-x}$ k times and then multiplying the result by e^x

$$L_k(x) = e^x \frac{d^k(x^k e^{-x})}{dx^k} \quad 3$$

An associated Laguerre polynomial, $L_k^p(x)$, is obtained by differentiating the corresponding Laguerre polynomial p times

$$L_k^p(x) = \frac{d^p L_k(x)}{dx^p} \quad 4$$

In wave-mechanics, associated Laguerre polynomials occur in solutions of the radial part of the wave-equation, $R_{n,l}(r)$, in which they take the form $L_{n+l}^{2l+1}\left(\frac{2Zr}{n}\right)$, where r is measured in terms of the atomic unit of length

$$a_0 = \epsilon_0 h^2/(\pi m_e^2) = 52.9 \text{ pm}.$$

The quantum number n is required in the radial equation to prevent an infinity occurring in the solution; it can be any positive integer. As the highest power of x in $L_k(x)$ is x^k, so by equation 4 p must be less than or equal to k for $L_k^p(x)$ to be non-zero. Hence $(2l+1) \leqslant (n+l)$ or $l \leqslant (n-1)$.

The solutions to the three separated wave-functions are still arbitrary to the extent of a multiplicative constant. The values of these constants can be calculated if ψ is interpreted to be a function such that $\psi^2 dv$ at any point in space gives the probability of finding the electron in a small element of volume dv at that point. As the electron must be somewhere, the probability of finding it in the whole of space is just 1. Hence we must put $\int_{\text{all space}} \psi^2 dv = 1$. This process is called normalization. As ψ is the product of three functions, the simplest procedure is to normalize each separately. When this is done the explicit expressions for the three solutions become

$$R_{n,l}(r) = -\sqrt{\frac{4(n-l-1)!Z^3}{n^4[(n+l)!]^3}} \left(\frac{2Zr}{n}\right)^l L_{n+l}^{2l+1}\left(\frac{2Zr}{n}\right) e^{-Zr/n} \quad 5$$

$$\Theta_{l,m}(\theta) = \sqrt{\frac{(2l+1)(l-|m|)!}{2(l+|m|)!}} P^{|m|}(\cos\theta) \quad 6$$

$$\Phi_m(\phi) = \frac{1}{\sqrt{(2\pi)}} e^{im\phi} \quad 7$$

These general expressions for the three functions are rather complicated but they reduce to very simple expressions for the small values of n, l and m which occur in chemical systems, as can be seen by reference to Tables 1–3 (pp. 7 and 10).

Appendix 2. Slater Functions

Self-consistent field calculations (p. 18) indicate that nodes in the radial functions of orbitals in a many-electron atom occur well in towards the nucleus. If they are neglected entirely and the asymptotic (exponential) form of the radial function at large distances is used, then the radial part of a given orbital can be approximated by the function[12]

$$R^*(r) = r^{n^*-1}\,e^{-\alpha r}$$

where r is in atomic units of a_0 (= 52.9 pm) and n^* and α are parameters adjusted to get agreement with the self-consistent field values of $R'(r)$. The parameter n^* can be thought of as an 'effective principal quantum number' and is given by

$$n\ = 1,\quad 2,\quad 3,\quad 4,\quad 5,\quad 6$$
$$n^* = 1,\quad 2,\quad 3,\quad 3.7,\quad 4,\quad 4.2$$

The parameter α is related to the 'effective nuclear charge', Z^*, by the relation $\alpha = Z^*/n^*$ where $Z^* = (Z-S)$ and S is a 'shielding constant'.

The shielding constant is found by the following empirical rules which are due to Slater.

(1) Divide the orbitals into groups, each group having a different shielding constant. This is done by writing down the orbitals in sequence of increasing principal quantum number and putting s and p orbitals in the same group but keeping d orbitals and f orbitals separate:

1s; 2s,2p; 3s,3p; 3d; 4s,4p; 4d; 4f; 5s,5p; *etc.*

(2) The shielding constant, S, is then the sum of the following contributions:

(*a*) nothing from any group occurring after the one considered;

(*b*) 0.35 for each other electron in the group considered, except in the 1s group where 0.30 is used;

(*c*) if the electron considered is in an sp group, 0.85 for each electron with principal quantum number less by one, and then 1.00 for all electrons further in;

(*d*) if the electron considered is in a d or f group, 1.00 for all electrons inside the group, including the immediately preceding one. Consider, for example, gallium: $1s^2$; $2s^22p^6$; $3s^23p^6$; $3d^{10}$; $4s^24p$. The shielding constant for the 4p electron is

$$S = [(2\times0.35)+(18\times0.85)+(10\times1.00)] = 26.0$$

Hence, $Z^* = Z-S = 31-26.0 = 5.0$

and $\alpha = Z^*/n^* = 5.0/3.7 = 1.35$

The functions are usually normalized after the angular part of the wave-function (Table 4, p. 13) has been included. The full Slater orbitals then take the form given in Table 9. By differentiation it is seen that the maximum in the radial function occurs at $r_{max} = (n^*-1)/\alpha$. In the development of valency theory r_{max} is taken as a rough criterion of orbital size compatibility.[13]

Table 9. Normalized Slater orbitals

$$\psi_{1s} = \sqrt{\frac{\alpha^3}{\pi}}\, e^{-\alpha r} \qquad \psi_{2p_x} = \sqrt{\frac{\alpha^5}{\pi}}\, r \sin\theta \cos\phi\, e^{-\alpha r}$$

$$\psi_{2s} = \sqrt{\frac{\alpha^5}{3\pi}}\, re^{-\alpha r} \qquad \psi_{2p_z} = \sqrt{\frac{\alpha^5}{\pi}}\, r \cos\theta\, e^{-\alpha r}$$

$$\psi_{3s} = \sqrt{\frac{2\alpha^7}{45\pi}}\, r^2 e^{-\alpha r} \qquad \psi_{3p_z} = \sqrt{\frac{\alpha^7}{15\pi}}\, r^2 \cos\theta\, e^{-\alpha r}$$

$$\psi_{4s} = \sqrt{\frac{\alpha^9}{315\pi}}\, r^{2.7} e^{-\alpha r} \qquad \psi_{4p_z} = \sqrt{\frac{\alpha^9}{105\pi}}\, r^{2.7} \cos\theta\, e^{-\alpha r}$$

$$\psi_{3d_{z^2}} = \sqrt{\frac{\alpha^7}{18\pi}}\, r^2 (3\cos^2\theta - 1)\, e^{-\alpha r}$$

$$\psi_{3d_{zx}} = \sqrt{\frac{2\alpha^7}{3\pi}}\, r^2 \cos\theta \sin\theta \cos\phi\, e^{-\alpha r}$$

$$\psi_{3d_{zy}} = \sqrt{\frac{2\alpha^7}{3\pi}}\, r^2 \cos\theta \sin\theta \sin\phi\, e^{-\alpha r}$$

$$\psi_{3d_{x^2-y^2}} = \sqrt{\frac{\alpha^7}{6\pi}}\, r^2 \sin^2\theta\, (2\cos^2\phi - 1)\, e^{-\alpha r}$$

$$\psi_{3d_{xy}} = \sqrt{\frac{2\alpha^7}{3\pi}}\, r^2 \sin^2\theta \sin\phi \cos\phi\, e^{-\alpha r}$$

$$\psi_{4d_{z^2}} = \sqrt{\frac{\alpha^9}{63\pi}}\, r^{2.7} (3\cos^2\theta - 1)\, e^{-\alpha r}$$

Slater functions reproduce well the x-ray spectra, ionization energies and energies of electrons in atoms. Thus, the total electronic energy of an atom is the sum of the quantities α^2 for each electron in the atom.

Slater functions contain no radial nodes and are therefore not orthogonal to each other, but they can be orthogonalized[14] and they then show radial nodes in good agreement with the positions indicated by the self-consistent field treatment. Slater functions are approximately correct in the outer parts of the electron distribution and are therefore satisfactory when one is interested only in this region of an atom (valency); they are less satisfactory for calculations dealing with the inner parts (*eg* quadrupole coupling constants). Furthermore no distinction is made between the shielding of electrons in s and p orbitals, though s orbitals are more penetrating and should therefore be less shielded than p orbitals.[15] The functions are quite accurate for the first three quantum shells but become less reliable when the fourth and higher shells are considered. They can be used for such orbitals, but more caution is required in drawing detailed conclusions.

Periodic Table of the Elements (including atomic numbers, chemical symbols, electronic structures and ionization energies in kJ mol−1)

1 H $1s^1$ 1312	2 He $1s^2$ 2371

3 Li $2s^1$ 520	4 Be $2s^2$ 899											5 B $2p^1$ 800	6 C $2p^2$ 1086	7 N $2p^3$ 1403	8 O $2p^4$ 1313	9 F $2p^5$ 1681	10 Ne $2p^6$ 2080
11 Na $3s^1$ 495	12 Mg $3s^2$ 738											13 Al $3p^1$ 577	14 Si $3p^2$ 786	15 P $3p^3$ 1018	16 S $3p^4$ 1000	17 Cl $3p^5$ 1255	18 Ar $3p^6$ 1520
19 K $4s^1$ 418	20 Ca $4s^2$ 590	21 Sc $3d^1 4s^2$ 633	22 Ti $3d^2 4s^2$ 659	23 V $3d^3 4s^2$ 650	24 Cr $3d^5 4s^1$ 653	25 Mn $3d^5 4s^2$ 717	26 Fe $3d^6 4s^2$ 762	27 Co $3d^7 4s^2$ 759	28 Ni $3d^8 4s^2$ 736	29 Cu $3d^{10} 4s^1$ 745	30 Zn $3d^{10} 4s^2$ 906	31 Ga $4p^1$ 579	32 Ge $4p^2$ 760	33 As $4p^3$ 946	34 Se $4p^4$ 941	35 Br $4p^5$ 1142	36 Kr $4p^6$ 1351
37 Rb $5s^1$ 403	38 Sr $5s^2$ 549	39 Y $4d^1 5s^2$ 615	40 Zr $4d^2 5s^2$ 659	41 Nb $4d^4 5s^1$ 664	42 Mo $4d^5 5s^1$ 688	43 Tc $4d^6 5s^1$ 697	44 Ru $4d^7 5s^1$ 710	45 Rh $4d^8 5s^1$ 720	46 Pd $4d^{10}$ 804	47 Ag $4d^{10} 5s^1$ 731	48 Cd $4d^{10} 5s^2$ 867	49 In $5p^1$ 558	50 Sn $5p^2$ 708	51 Sb $5p^3$ 833	52 Te $5p^4$ 869	53 I $5p^5$ 1007	54 Xe $5p^6$ 1170
55 Cs $6s^1$ 374	56 Ba $6s^2$ 502	57 La $5d^1 6s^2$ 541	72 Hf $5d^2 6s^2$ 531	73 Ta $5d^3 6s^2$ 760	74 W $5d^4 6s^2$ 770	75 Re $5d^5 6s^2$ 759	76 Os $5d^6 6s^2$ 839	77 Ir $5d^9$ 842	78 Pt $5d^9 6s^1$ 868	79 Au $5d^{10} 6s^1$ 890	80 Hg $5d^{10} 6s^2$ 1006	81 Tl $6p^1$ 589	82 Pb $6p^2$ 715	83 Bi $6p^3$ 703	84 Po $6p^4$ 813	85 At $6p^5$ 887	86 Rn $6p^6$ 1037
87 Fr $7s^1$ 384	88 Ra $7s^2$ 509	89 Ac $6d^1 7s^2$ 661															

58 Ce $4f^2 6s^2$ 667	59 Pr $4f^3 6s^2$ 556	60 Nd $4f^4 6s^2$ 608	61 Pm $4f^5 6s^2$ —	62 Sm $4f^6 6s^2$ 540	63 Eu $4f^7 6s^2$ 547	64 Gd $4f^7 5d^1 6s^2$ 594	65 Tb $4f^9 6s^2$ 650	66 Dy $4f^{10} 6s^2$ 658	67 Ho $4f^{11} 6s^2$ —	68 Er $4f^{12} 6s^2$ 586	69 Tm $4f^{13} 6s^2$ 590	70 Yb $4f^{14} 6s^2$ 600	71 Lu $4f^{14} 5d^1 6s^2$ 593
90 Th $6d^2 7s^2$ 669	91 Pa $5f^2 6d^1 7s^2$ —	92 U $5f^3 6d^1 7s^2$ 386	93 Np $5f^5 7s^2$ —	94 Pu $5f^6 7s^2$ —	95 Am $5f^7 7s^2$ —	96 Cm $5f^7 6d^1 7s^2$ —	97 Bk $5f^8 6d^1 7s^2$ —	98 Cf $5f^{10} 7s^2$ —	99 Es	100 Fm	101 Md	102 No	103 Lr

References

1 W. Kauzmann, *Quantum chemistry*. London and New York: Academic Press, 1957.

2 Y. K. Syrkin and M. E. Dyatkina, *Structure of molecules and the chemical bond*. London: Butterworths, 1950.

3 H. Eyring, J. Walter and G. E. Kimball, *Quantum chemistry*. London and New York: Wiley, 1944.

4 L. Pauling and E. B. Wilson, *Introduction to quantum mechanics*. New York and London: McGraw-Hill, 1935.

5 H. Margenau and G. M. Murphy, *The mathematics of physics and chemistry*. Second edition. London and New York: Van Nostrand, 1956.

6 D. R. Hartree, *Rep. Prog. Phys.*, 1946–7, **11**, 113.

7 D. R. Hartree, *The calculation of atomic structure*. London and New York: Wiley, 1957.

8 Landolt–Börnstein's Tables. Sixth edition. Vol. 1, part 1, 1950.

9 P. G. Dickens and J. W. Linnett, *Q. Rev. Chem. Soc.*, 1957, **11**, 291.

10 C. E. Moore, *NBS Circular 467*. Vol. 1, 1949; vol. 2, 1952; vol. 3, 1958.

11 N. N. Greenwood, *Educ. Chem.*, 1964, **1**, 176.

12 J. C. Slater, *Phys. Rev.*, 1947, **36**, 634.

13 D. P. Craig, A. Maccoll, R. S. Nyholm, L. E. Orgel and L. E. Sutton, *J. Chem. Soc.*, 1954, 332.

14 W. E. Moffitt and C. A. Coulson, *Philos. Mag.*, 1947, **38**, 634.

15 W. E. Duncanson and C. A. Coulson, *Proc. R. Soc. Edinburgh*, 1944, **62**, 37.

Suggestions for Further Reading

J. W. Linnett, *Wave mechanics and valency*. London: Methuen, 1960.

F. Seel, *Atomic structure and chemical bonding* (trans. N. N. Greenwood and H. P. Stadler). London: Methuen, 1963.

R. M. Hochstrasser, *Behaviour of electrons in atoms*. New York and Amsterdam: Benjamin, 1964.

H. B. Gray, *Electrons and chemical bonding*. New York: Benjamin, 1964.

M. W. Hanna, *Quantum mechanics in chemistry*. New York and Amsterdam: Benjamin, 1965.

J. N. Murrell, S. F. A. Kettle and J. M. Tedder, *Valence theory*. London and New York: Wiley, 1965.

P. G. Perkins, *Elementary molecular bonding theory*. London: Methuen, 1969.

S. R. LaPaglia, *Introductory quantum chemistry*. London: Harper & Row, 1971.

J. C. Schug, *Introduction to quantum chemistry*. New York: Holt, Rinehart, and Winston, 1972.

I. N. Levine, *Quantum chemistry*, 2nd edn. Boston: Allyn and Bacon, 1974.

P. Hedvig, *Experimental quantum chemistry*. New York: Academic Press, 1975.

A. R. Denaro, *A foundation for quantum chemistry*. London: Butterworths, 1975.

MONOGRAPHS FOR TEACHERS

This series was launched, by the Royal Institute of Chemistry, in 1959 with the aim of presenting concise and authoritative accounts of selected well-defined topics in chemistry for the guidance of those who teach the subject at GCE Advanced Level and above. Though intended primarily for teachers of chemistry, these Monographs have proved of value to a wider readership, including more advanced students. Up to three new titles are added to the series each year. Members of the Chemical Society ordering *Monographs for Teachers* may claim a discount of 25 per cent off the published prices.

'.. full of excellent illustrative material, references to the literature, and potential sources of problems of interest to students and teachers.'

—*Journal of Chemical Education*

'Few teachers can afford to be without a complete set on their bookshelves.'

—*Chemistry and Industry*

TITLES IN PRINT

2. PRINCIPLES OF OXIDATION AND REDUCTION. A. G. SHARPE. (SECOND (SI) EDITION).
4. PRINCIPLES OF METALLIC CORROSION. J. P. CHILTON. (SECOND (SI) EDITION).
7. PRINCIPLES OF CATALYSIS. G. C. BOND. (REVISED SECOND EDITION).
8. PRINCIPLES OF ATOMIC ORBITALS. N. N. GREENWOOD. (REVISED (SI) EDITION).
9. PRINCIPLES OF REACTION KINETICS. P. G. ASHMORE. (REVISED SECOND EDITION).
13. PRINCIPLES OF OSMOTIC PHENOMENA. J. F. THAIN.
15. PHYSICOCHEMICAL QUANTITIES AND UNITS. M. L. McGLASHAN. (SECOND EDITION)
16. CHEMICAL PROCESSING IN INDUSTRY. M. D. WYNNE.
17. AN INTRODUCTION TO BIOCHEMISTRY. E. G. BROWN.
18. PRINCIPLES OF CRYSTAL CHEMISTRY. E. CARTMELL.
19. THE MOLECULAR BASIS OF ENTROPY AND CHEMICAL EQUILIBRIUM. P. A. H. WYATT.
20. SILICON CHEMISTRY AND APPLICATIONS. C. A. PEARCE.
21. MODERN ANALYTICAL METHODS. D. BETTERIDGE AND H. E. HALLAM.
22. AN INTRODUCTION TO PHOTOCHEMISTRY. P. SUPPAN.
23. CHEMICAL ASPECTS OF THE ATOMIC NUCLEUS. J. G. CUNINGHAME.
24. PRINCIPLES OF FREE RADICAL CHEMISTRY. J. I. G. CADOGAN.
25. SOME ASPECTS OF TECHNOLOGICAL ECONOMICS. D. F. BALL.
26. ELEMENTS OF ORGANOMETALLIC CHEMISTRY. F. R. HARTLEY.
27. THE HYDROGEN BOND AND OTHER INTERMOLECULAR FORCES. J. CLARE SPEAKMAN.
28. SOME ASPECTS OF BASIC POLYMER SCIENCE. D. A. BLACKADDER.
29. ION-EXCHANGE: INTRODUCTION TO THEORY AND PRACTICE. R. W. GRIMSHAW AND C. E. HARLAND
30. MOLECULAR STRUCTURE: ITS STUDY BY CRYSTAL DIFFRACTION. J. CLARE SPEAKMAN.
31. THE PRINCIPLES OF BIO-INORGANIC CHEMISTRY. A. M. FIABANE AND D. R. WILLIAMS.
32. AN INTRODUCTION TO ENZYME CHEMISTRY. P. F. LEADLAY
33. INORGANIC REACTION MECHANISMS. D. O. COOKE

Further information on all these publications from:

The Marketing Officer, The Chemical Society, Burlington House, London W1V 0BN

Principles of atomic orbitals describes the electronic structure of atoms in terms of concepts which are directly useful to chemists. SI units are used throughout. Mathematical formalism is minimized and all results are described both verbally and pictorially. After a brief introduction which mentions approaches other than that of the Schrödinger wave equation, the solution of this equation for the unique case of hydrogen-like atoms is considered in detail. Special care is taken to explain the meaning of the various terms used and common misconceptions are avoided; in particular the distinction between 'orbitals' and 'angular dependence functions' is stressed.

The complications which arise in the case of many-electron atoms are then considered and the influence that these have on the electronic energy levels is described in detail. The concepts of the self-consistent field, electron penetration, and orbital energy are introduced at this stage and the significance of the resultant quantum numbers and of Hund's rules is also discussed. The final chapter concerns the periodic classification of the elements and considers the electronic structure of each element in terms which emphasize trends and correlation of significance in chemistry. Appendices contain notes on the formal solution of the wave equation and on the approximation to atomic orbitals given by the Slater functions. The Monograph concludes with suggestions for further reading.

Norman Greenwood is Head of the Department of Inorganic and Structural Chemistry in the University of Leeds. He graduated from the University of Melbourne and then went to Cambridge for his PhD (1951) before being appointed successively as Senior Research Fellow at AERE Harwell, Lecturer at the University of Nottingham and Professor of Inorganic Chemistry at Newcastle upon Tyne. He moved to his present post in 1971. Current research interests centre on synthetic and structural chemistry of boron hydride derivatives and the application of physical techniques such as Mössbauer spectroscopy to problems in chemistry. He is author or co-author of over 300 original papers and reviews.

Professor Greenwood has always been interested in the teaching of chemistry. He was Resident Tutor at Trinity College Melbourne, Founder Chairman of the Chemistry Teachers' Centre in Newcastle upon Tyne and a member of the British Committee on Chemical Education. He has been Moderator and Part II Examiner for GradRIC, Deputy Chairman of the Chemical Society's Education and Training Board and Chairman of the Working Party on Careers with Chemistry. He has also held Visiting Professorships in Denmark, Australia, Canada and the U.S.A., and is at present President of the Dalton Division of the Chemical Society and President of the Inorganic Chemistry Division of I.U.P.A.C. Several of the nine books which he has written or co-authored have been for school pupils, chemistry

9780854040285